Alistair Moffat was born and bred in the Scottish Borders and studied at the universities of St Andrews, Edinburgh and London. A former Director of the Edinburgh Festival Fringe, Director of Programmes at Scottish Television and founder of the Borders Book Festival, he is also the author of a number of highly acclaimed books. From 2011 to 2014 he was Rector of the University of St Andrews.

THE
SCOTS
A GENETIC JOURNEY

ALISTAIR
MOFFAT

BIRLINN

*To my friend Francis Hamilton, a man
who knows what's what and who's who*

This edition published in 2017 by
Birlinn Limited
West Newington House
10 Newington Road
Edinburgh
EH9 1QS

First published in 2011

www.birlinn.co.uk

ISBN: 978 1 78027 444 7

British Library Cataloguing-in-Publication Data
A catalogue record for this book is available from the British Library

Designed and typeset by Iolaire Typesetting, Newtonmore
Printed and bound by Grafica Veneta, Italy.

Contents

✱

List of Plates and Illustrations

✳

List of Plates

The crowd rejoices as Scotland wins a Grand Slam at Murray-field in 1990

Wild cattle, horses and deer thunder across the walls of the cave at Lascaux

What Ötzi might have looked like

A hunter-gatherer-fisher harpoon from the island of Oronsay

Sir William Jones, the great linguist

A pottery vessel from the henge at Balfarg

The recumbent stone circle at East Aquhorthies in Aberdeenshire

The interior of the chambered cairn at Maes Howe, Orkney

The mound of Maes Howe, built some time around 3,000 BC

The centre of the standing stones at Callanish, Isle of Lewis

Grave goods found at Culduthel, Inverness

A mummified body found at Cladh Hallan, South Uist

The hundreds of hut platforms on Eildon Hill North, Scottish Borders

The Aberlemno Stone, showing the Battle of Dunnichen, 685

The magnificent Ruthwell Cross

Viking runic graffiti in Maeshowe, Orkney

Emigrants leaving the Isle of Lewis

A poster advertising the virtues of emigration

Another poster worrying about the arrival of giants from the
Hebrides

An Italian ice-cream shop in early twentieth-century Edinburgh

A family of Lithuanian immigrants pose for posterity

Pakistani girls playing in a Glasgow street in the 1960s

DNA being rehydrated on rotating wheels

List of Illustrations

Note to the New Edition

✖

Ancestral DNA is a fast-moving branch of knowledge and its links with more traditional historiography are still developing as each discipline informs the other. This new edition of *The Scots: A Genetic Journey* includes recent findings on early farming and its migrations, Neanderthal DNA, the classic Celtic DNA Y chromosome haplogroup, the royal Stewart lineages, and physical characteristics such as red hair and blue eyes. With every passing year, the picture painted by this new fusion becomes clearer, more detailed and brighter.

Alistair Moffat
March 2017

Introduction

The Crowd

✲

EVERY YEAR, IN THE dowie days of February and early March, Scotland comes to Edinburgh. From the platforms of Waverley and Haymarket Stations, from buses parked at King's Stables Road, from cars left in the western suburbs, an army musters. The yellow of the Lion Rampant and the blue of the Saltire standards flutter as tens of thousands march west out of the centre of the city, flooding like a tartan tide along Shandwick Place, West Maitland Street, past Donaldson's Hospital to Roseburn and the old battleground.

To join the national muster Northerners come down on the morning trains from Inverness and Aberdeen. From the west, Argyll, Ayrshire and Glasgow, they rattle through in commuter carriages or join the high-speed queue that is the M8. From the Borders, once the heartland, they arrive in their thousands. It is the journey of an army of hope that often ends in despairing, rancorous retreat. It is Scotland playing rugby at Murrayfield.

The sea of expectant faces, 67,000 strong, most of them gasping with exasperation or shouting encouragement to the team wearing navy blue, knows that Scotland can beat any team in the world, except New Zealand, and can lose to any team in the world. But they glory in their day and its lurching swings of emotion, the crowd a counterpoint to the play on the pitch. All follow the unscripted drama, pause for breath at stoppages, roar at breaks and thrusts, groan at breakdowns and mistakes. For an afternoon, Scotland, Scottishness and being Scots behave like an organism, like an unwitting microcosm of a nation.

In many important ways, the Murrayfield crowd is just that. Allowing for the supporters of visiting teams, it represents 1 per cent of the population. The stands are filled with men, women and children from the Highlands, Aberdeenshire and Tayside, Perthshire and Fife, from the Clyde coastlands, from Galloway and the Borders and the cities of Scotland. The days of a male-dominated rugby crowd are long gone now and, while women may not make up 50 per cent, the proportion is large.

What superficially binds and identifies the crowd are their iconography and a sense of shared history. Kilts, flags worn as capes, face paints, See-You-Jimmy hats and national rugby jerseys are part of the carnival uniform. And, when Scotland needs slightly different sorts of exhortation, the team is advised to 'Remember Bannockburn' or, alternatively, 'Remember Flodden'. Before hostilities begin, 'Flower of Scotland' is sung to a dirge-like tune. The anthem recalls the triumph at Bannockburn as though it were a matter for mourning. Proud Edward's army is defeated and, in an unlikely anachronism, Robert the Bruce is said to have 'sent him homeward tae think again'.

Kings and battles, flags and tartan – these appear to be the stuff of history for the crowd. A procession of armies across Scotland's landscape, the drama of warfare and raid, picturesque ruins, important dates are all lines in a familiar recital. Until the nineteenth century, the story of Scotland circled around the doings and decisions of elites, of tiny groups of powerful and interconnected people. Monarchs, aristocrats and churchmen actually *made* history or at least had control over its recording. Over most of the vast span of all that experience in what is now Scotland, there was simply no place for the stories of ordinary lives, of the agricultural labourers whose back-breaking toil formed the landscape, the tradesmen whose skills made the work easier and the masons who built the great churches, castles and grand houses.

The Murrayfield crowd cannot know anything about these unre-counted generations except for one simple thing – that they were their ancestors. And that irreducible fact is the rock upon which this book and all the research behind it is built. As the tramp of invading armies echoes and fades and in the long shadows behind the stately line of kings and noblemen and their stately homes, we may now at last be able to make out the modest outline of a people's history of Scotland – our people, ourselves, our history, a story of Scotland we can all own. The Murrayfield crowd hold it inside themselves – a long-hidden story of unlikely and unsurprising links, a story that reaches back beyond kings and queens to the very beginnings of settlement in Scotland after the end of the last Ice Age, a story that is more than 11,000 years old. It is the story of our DNA.

Every Scot is an immigrant. Until 9,000 BC, Scotland was empty of people and animals. For 15,000 years, ice, more than a kilometre thick in places, had crushed the land under a pitiless white sterility where nothing could live. When the ice sheets finally retreated, small bands of pioneers moved north into a virgin landscape. Their names are not known but scientists have been able to discover a great deal about where they came from. Immigrants arrived in what would become Scotland over thousands of years, well into the historical period. And their descendants are still here, flourishing, carrying on the long journey of their genes. By approximately AD 1,000, the process was almost complete and most of the pieces of our patchwork of DNA were in place. Significant but small groups did come to Scotland after 1,000 but, in statistical terms, their impact was minimal.

That is why the balance of this book is heavily tilted towards prehistoric and early historic times. DNA research works best in a chronological context and that is the reason for the first landmarks in our history being clearly plotted in this story – although some of these turn out to be misleading.

Language and its development are closely linked to who we are and where we came from and the evolution of Old Welsh, Scots English and Gaelic is an important theme. There exist fascinating linguistic links and many of them were forged in deep time, tens of thousands of years ago as our ancestors crossed seas and rivers to walk into empty landscapes and populate the Earth.

As well as Y chromosomes, men inherit surnames and, as these came increasingly into common use in the early modern period, they help set genetic findings against a familiar background. For example, Scotland's Highland clans often claim descent from a common name-father but recent research shows that some of these men may have been fathers in more than name only.

None of what follows would have been possible without the pioneering work of two scientists working in the universities of California. In the 1960s, Professor Luca Cavalli-Sforza was the first to apply genetics to history and he used the first DNA markers to do this. These were the familiar blood groups – ABO, rhesus and so on. Earlier research had shown that these differed between populations across the world and, using these findings, Cavalli-Sforza built a version of a family tree that showed how populations were related to each other. What puzzled him and others was the deep differences between populations in Africa.

A New Zealander of Scots descent, Professor Allan Wilson, began to study mitochondrial DNA, what women pass on through the female line to their daughters and to their sons (but it dies with them). He saw how individuals, and not just populations, were related to each other through their DNA. And, because these branches of connection were longest in Africa, Wilson and his team made the earth-shattering announcement that the human race had originated there.

This caused uproar. Most scientists believed that *Homo sapiens* had descended from various ancestors around the world – the Chinese were thought to be the children of Peking Man, the South-East Asians came from Java Man and Europeans from Neanderthals. The discovery that modern humans had walked out of Africa to populate the whole of the rest of the world was sensational and it made headlines. Images of a real Garden of Eden suddenly came into focus and Mitochondrial Eve was pictured on the cover of *Time Magazine*.

Research began to gather pace. After a period of emphasis on mitochondrial DNA, Y chromosome evidence was analysed and hundreds of new markers identified in the 1990s. Dating techniques improved and some European lineages were shown to originate before the last Ice Age, while others had been brought into Europe by early farmers from the Middle East.

At the same time, analysing DNA samples became simpler, faster and cheaper. All individuals had to do was to spit into a plastic capsule (a 'kissing spit' is advised as being sufficient and a hawking spit is to be avoided – for all sorts of reasons), seal it and send it off to be processed. Very quickly, there were companies that, for a fee, would test DNA and explain it and the results were uploaded to a series of databases. Worldwide there are probably around 200,000 results for Y chromosome tests and in Britain about 20,000. These come from people anxious to trace their family trees. They are large and reliable samples and they are being constantly augmented and refined. Once a geneticist has the results of a DNA test, he looks at the database for other markers of the same sort. Many of these have locations or origins attached and also other important genealogical information. This allows a context for new samples and makes all sorts of fascinating links, both geographical and historical.

One of the most striking findings in this flurry of recent research is the story of the epic journey of our ancestors out of Africa. Whoever the Scots believe themselves to be now, they are descended from Africans. The same is true for the English, the Irish, the Russians, the Chinese – all of the rest of the peoples of the world. It is a wonderfully refreshing, ironic and redressing balance for centuries of racial prejudice to think that *Homo sapiens*, and not-so-sapiens, originated amongst people once routinely and widely believed to be sub-human. Instead, it is clear that Africans were once our mothers and fathers.

At the other end of the chronology of DNA history, my own experience as a child touched on the insistent theme of this book, that of immigration. At school in Kelso I sat next to Richard Mazur and Jot Wichary. I knew Denis and Brian Poloczek and Harry Tomczek, and admired Leona Goldsztajn from, sadly, a distance. Their DNA and that of the children of other Polish exiles arrived in Scotland because of an accident of history and, as a further consequence of happenstance, was concentrated in Kelso. It now enriches Scotland and makes only one of thousands of links in the chain of stories that follows.

Alistair Moffat
St Andrew's Day, 2010

The Refuges

✳

A T THE TIPS OF ITS wingspread, the eagle's feathers fluttered in the warm updraft. Stalling and banking in the breeze, it suddenly, effortlessly soared high over the sunlit Clyde Valley. Through the clear air, the great bird flicked its head from side to side, looking down on an endless vista of rolling green grasslands watered by the river and its tributaries. Glinting in the sun, hundreds of small pools patterned the open landscape.

Its attention caught by sudden movement, the eagle turned and spiralled downward to where three pools clustered. Their surfaces quivered as thunder stampeded past their shores. Trumpeting, their tusks swaying, a herd of mammoths crashed through the grassland, fleeing headlong from danger. Behind them cave lions had brought down a calf and, as the dying animal thrashed, hooked tusks slashing in the air, its killers dug their incisors into its throat, choking out the life. The eagle could see that there would be no pickings later for a pack of hyenas was circling, waiting for the lions to be gorged and sated.

Elsewhere on the treeless plains, other prey grazed. Vast herds of wild horses ranged over wide pasture. With their lookout stallions skirting the fringes of the herds and the dominant mares keeping them together, the small horses could be difficult to bring down. Fast and wary, they could see or sniff trouble before it was upon them. But, when mares broke off to drop their foals, usually at the dead of night, packs of wolves could pick them off. In the morning, there would be something left for a sharp-eyed eagle.

Few predators were bold enough to take on the hard-charging woolly rhinoceros and the brown bears, bigger than grizzlies, that roamed the uplands. Through its long life in the skies above the grasslands the eagle had seen only one predator with the guile, strength and determination to attack any animal, no matter how fast, large or ferocious.

Three thousand generations ago, some time around 58,000 BC, men and women walked north into Britain. At that time, it was the far north-western peninsula of the European landmass. So much ice had formed around the poles and over Scandinavia that the level of the sea had fallen between 160 and 260 feet below modern norms. The weather was cold all year round but dry and few trees grew on the peninsula. But the rich summer grasslands drew herd animals northwards and behind the reindeer, the wild horses and the thundering herds of mammoths came small bands of hunters. They were human beings but they were not like us.

In 1856, near Dusseldorf in Germany, miners were quarrying a limestone canyon when they opened out a cave mouth. Inside lay a scatter of bones, parts of a skull, arm and leg bones and ribs. They were very thick and, at first, the miners believed they had come across the skeleton of a cave bear. When an amateur naturalist, Johann Carl Fuhlrott, and an anatomist, Hermann Schaaffhausen, saw the bones they realised immediately that they could only be human but not because they were exactly like those of our subspecies, *Homo sapiens* Skeletal remains found all over Europe have allowed a clear picture of the Neanderthals to be pieced together. With deep barrel chests, stout bones and tremendous musculature, their stooped bodies were well adapted to cold climates. A reduced surface area prevented heat loss and it seems likely that their bodies were comparatively hairy. Neanderthal skulls show a massive brow ridge, a receding chin with prominent cheekbones and a larger brain case than *Homo sapiens*. Scientists believe that their big brains may have enabled greatly enhanced night vision, a valuable attribute for a society based almost exclusively on hunting and gathering.

2

Neanderthal noses were large and flat, able to inhale and exhale a greater capacity. It used to be thought that this was another feature of a cold-adapted physique with a bigger nose warming freezing air a little to prevent it shocking the lungs. But recent research promotes a more likely and more elegant thesis. For very stocky bodies, exertion and a consequent rise in temperature could be difficult and it may well be that the Neanderthals' noses acted as efficient heat exchangers helping them cool down quickly.

For skilled scientists, skeletons can be eloquent, offering a sure guide to living musculature, and it is possible to reconstruct accurately not only what these men and women looked like but also how they used their bodies. Their powerful build speaks of a life of tremendous exertion, especially in their style of hunting. A Neanderthal was much stronger than the strongest *Homo sapiens* and was able to sprint at lightning speed across short distances. They needed to. Tools found at open-air sites and caves in southern England show some sophistication but Neanderthals knew nothing of bows and arrows. They needed to get closer to their prey. Most of the stone implements found were for butchering carcases and scraping hides and bones, and it seems that their spears were tipped only with sharp flakes of flint.

Injuries deduced from skeletons suggest a dangerous, even ferocious approach to hunting. Neanderthals suffered severe wounds to their heads and upper bodies and some even survived arm and leg breaks. One imaginative historian has noticed a parallel pattern of injuries to modern rodeo riders. It seems likely that Neanderthals attacked their prey directly, launching themselves at reindeer or wild horses, perhaps jumping on their backs to bring them down. Perhaps they even took on mammoths. Their weapons were almost certainly good enough to wound an animal and hunters may have stalked to get close enough and then sprinted to attack and stab. Then they chased whatever they had hit. When blood loss had weakened, say, a horse, the Neanderthals brought it down through sheer speed and brute strength. As a dying animal kicked and thrashed, it could cause terrible wounds to its predator.

The sense of these prehistoric hunters as ferocious, attacking the prey of the grasslands with a savagery that seems barely human, is reinforced by closer examination of their skeletons. Even in identifiably younger skulls, the teeth of Neanderthals show tremendous wear and it seems that they used their mouths like a tool or a third hand. It may be that these people chewed hide to soften it and make it more workable for clothing or other domestic purposes. Their front teeth are much larger than those of *Homo sapiens* and curved so that they can grip as well as bite. Perhaps Neanderthal jaws were used to hold items while their hands worked on them – or perhaps they used their teeth as a weapon.

Comparative studies have revealed that Neanderthal children probably matured faster than those of *Homo sapiens*. And these prehistoric people also died earlier, rarely living beyond their thirties. But the impression of a feral subspecies needs substantial qualification. Neanderthals cared about their dead and buried them in what looks like a ritual manner, composing the corpse in a foetal position before covering it over. Cooperative activity such as a drive of mammoths and woolly rhinos over a cliff in what is now the island of Jersey suggests that they could plan, think consequentially and work in groups. That in turn suggests that they could articulate reasonably sophisticated speech and examination of Neanderthal skulls shows their voice box was in the same place as *Homo sapiens*' and therefore capable of the same range. It also appears that their culture had some abstract sense of decoration. Beads made by Neanderthals have been found and beautiful seashells collected for, it seems, no other purpose than the pleasure of looking at them and owning them.

Across the south of England, archaeologists have found evidence for the activities of Neanderthals after 58,000 BC but few skeletal remains. There is no doubt that, by this time, the herd animals at the centre of these people's lives had migrated when the climate warmed and the grass grew. But the hunters moved in tiny groups and it may be that no more than 20 or 30 families inhabited prehistoric Britain 3,000 generations ago. That extraordinary scarceness makes it very difficult for archaeologists to discover more.

Traces of the megafauna of the grasslands have been found in Scotland – mammoth tusks have come to light in a dig near Kilmarnock, on the watershed hills between the Ayrshire coastal plain and the Clyde Valley. Perhaps it was killed by cave lions. But nothing of the northern forays of pioneering Neanderthal hunting bands has been found even though it seems certain that they followed the herds as they sought fresh pasture. No characteristic assemblages of their tools have yet been uncovered, and what happened after the time of the mammoths, hyenas and lions in Scotland erased almost everything and makes it highly unlikely that any will ever be recognised.

After 24,000 BC the skies over Britain began to darken, storms blew and the weather quickly grew colder. Winter snow stayed on the hilltops throughout the year and the summers shortened. New grass came through later and later and began to die away in the early autumn. As rain and sleet fell, the great herds were driven southwards and, when the cold gripped the land, their human predators fled with them. The last Ice Age was beginning.

The causes of this drastic episode of climate change originated far out in space. For reasons related to the fluctuating gravitational pull of our Sun and the planets, the Earth's orbit changed. For long periods in the last 500,000 years, it has revolved around the Sun in an ellipse and not a near circle, as it does now. At the extremes of an elliptical orbit the Earth is further from its source of heat and light and therefore temperatures drop dramatically. Added to this cycle of change is another deadly complication. Over the last half million years, the angle of the Earth's axis has also altered, tilting the northern hemisphere away from the Sun's rays. This has the effect of shifting the Arctic Circle several hundred miles southwards. In turn this depresses temperatures even more as the cold waters of the north Atlantic began to turn away the warmth of the Gulf Stream.

Over northern Europe, the ice crept down the hillsides and covered the frozen land. As hurricanes blew and storms raged, more snow fell and it reflected the Sun's rays back, insulating the ice beneath, and temperatures dropped ever lower. Across the highest points of the ice sheet that had formed over Scotland, it was an average of minus 60 degrees Celsius.

Nothing could live. At the height of the last Ice Age, the ice sheet over Scotland was more than a mile thick and it extended far to the south, to a diagonal line from the South Wales coast over to the Humber Estuary and a small enclave immediately to the north. It was a pitiless and utterly sterile landscape. As air flowed down from the summit of the ice dome over Ben Nevis and Rannoch Moor, tremendous winds blew, their extraordinary speeds enhanced by the lack of friction from the ice. The near-constant high pressure produced clear skies and endless white vistas of dazzling but devastating beauty.

To the south of the sheet, conditions were different but equally extreme. At the fringes of the ice, where the glaciers had begun to fray and crumble, low pressure fronts travelled along the cliffs at the edge of the sheet and they brought sleet, cloud and very stormy conditions. Over all, the seasons' temperatures averaged minus 10 degrees. Below the line from South Wales to the Humber, southern England was a polar desert and had been entirely abandoned by animals and people. So much water had been locked up in the ice that global sea levels were lowered by 125 metres and those fleeing frozen Britain could walk far to the warmer south. A broad region of steppe-tundra covered northern Europe and it was only south of the Loire Valley that refugees could travel over grassland and see the welcome herds of grazing animals. Those northern Neanderthal bands who sought refuge in the south also came across another species.

Around 40,000 BC *Homo sapiens* arrived in Europe, his ancestors having migrated out of eastern Africa 30,000 years before. Although slightly taller, these new people looked like modern humans and we are their descendants. They spoke a developed language, possessed many skills and, as family bands crossed Europe from east to west, they thrived. Neanderthals had been hunting the grasslands and the forests for at least 100,000 years before and it seems that, for a period, both subspecies co-existed. Numbers were very small – only 5,000 or so people hunted in the whole of the area of modern France – and contact was probably rare. Europe was dominated not by people but by animals, the huge herds and the megafauna of the time before the onset of the last Ice Age.

Spearheaded by the panzer corps led by Generals Guderian and Rommel, German armies smashed into northern France in May

1940. Refuelling at roadside petrol stations, ignoring orders from Oberkommando der Wehrmacht (OKW) General Staff to halt and consolidate, the panzers dashed for the sea and the encirclement of British and French forces. Stukas dive-bombed a smoking path through French defences and, after the miraculous evacuation from Dunkirk, France fell, its armed forces surrendered and a humiliating armistice was signed in the forest at Compiègne. By 28 June, Adolf Hitler was sightseeing in Paris, triumphant, the master of Western Europe.

Southern and eastern France remained unoccupied and under the control of the collaborationist Vichy Regime of Marshal Pétain. Throughout the summer of 1940 the sun shone and life in the south appeared to go on much as before, although many must have been fearful of the future under a government absolutely controlled by the Nazis. But for young people, these concerns were perhaps less immediate.

On Sunday 8 September 1940 another sunny day dawned and it was buried treasure and not the war in the north that was much on the mind of 17-year-old Marcel Ravidat. An apprentice garage mechanic in the small town of Montignac in the steep-sided valley of the River Vézére, a tributary of the Dordogne in the Périgord region of central France, he had heard stories of a mysterious hidden cavern in the wooded hillside at Lascaux. With three friends and his dog, Robot, Marcel walked out of the town on a road leading to an abandoned chateau. Older people had said there was a secret passageway under the house which led to a cavern, a place where treasure had been buried. During the Revolution a priest known as Labrousse had fled the Terror and the guillotine by seeking refuge in the secret cave. Because the chateau had belonged to his family, he knew where to find the entrance.

The steep hillside behind the old house had been planted with trees and allowed to revert to wildwood. In the winter of 1920/21 a great storm had blown through the valley of the Vézére and brought down a mature pine tree. Its roots torn out by the force of the wind, the pine had left a surprising large hole and to prevent browsing animals from falling in, farmers had filled it with earth and rubbish.

After a fruitless few hours thrashing through the wildwood Marcel and his friends had found nothing and began to make their way down the hillside back home to Montignac. However, Robot was nowhere to be seen. Despite repeated calls and cajolements Marcel could not bring his dog to heel and the exasperated friends set off to look for him. Behind a tangle of brambles and scrub, Robot was found digging furiously, earth flying out behind the little dog. He was at the bottom of the gaping hole left by the windblown pine tree and when Marcel scrambled down to pull Robot away, he saw that the determined dog had dug another hole. Only six inches across, it seemed to lead into total blackness. Excited and intrigued, Marcel dropped some stones into the opening – and they clattered away, echoing into what seemed like endless emptiness. The boys looked at each other. Had Robot found the entrance to the secret cavern of Lascaux? Was there treasure down there? They resolved to say nothing to anyone and to come back another day with ropes, lanterns and shovels.

Equipped with a crude lamp and a long knife-like implement he had made himself, Marcel returned on 12 September to try to widen the hole. Robot had been left at home – this was serious business. Having brushed through the brambles and scrambled down to where the dog had been digging, Marcel soon reamed out enough to allow him to squeeze through. Clutching his lamp, he went in headfirst and promptly slid downwards into the darkness for about 20 feet. Unhurt, he managed to stand up and light the lamp – and then slipped and fell again. The lamp was snuffed out as Marcel was pitched forward. He could see nothing and must have been terrified, but his headlong descent was quickly broken by flat ground. Having had the presence of mind to hang onto his lamp and only scraped and bruised, he stood up and lit the lamp again. He could see that he was in a large cavern. In the darkness and silence around the boy astonishing figures were waiting to be revealed.

Once his friends had skidded down, Marcel led them into the heart of the cavern. Anxious not to fall again and with no idea of the gradient, he held the lamp at knee height. But when the boys shuffled slowly into a narrow corridor, they stopped on the firm ground and Marcel lifted up the lamp for the first time. Jacques Marsal gasped and cried out loud at what he suddenly saw. Galloping across the walls of the cavern were herds of wild horses and bulls. As the lamp flickered across the images, the animals seemed to move. And when the boys looked around they were open-mouthed, literally entranced. Much later Marcel recalled their reaction: 'a band of savages doing a war dance couldn't have done better'; while Marsal said, 'We were completely crazy'.

What little Robot had found was something remarkable, something that ranks as one of the wonders of the world. Inside the labyrinth of caves and caverns at Lascaux were almost 2,000 paintings of animals, most of them wild horses, stags, bison and at the entrance four huge black bulls. But what made these a wonder is that the menagerie of beautifully painted creatures was made some time around 15,000 BC, when most of Europe shivered in the grip of the last Ice Age.

This was treasure certainly but not what Marcel and his friends had in mind. The boys thought it best to tell their schoolteacher, M. Leon Laval, what they had found and he left a record:

> Once I arrived in the great hall accompanied by my young heroes, I uttered cries of admiration at the magnificent sight that met my eyes ... Thus I visited the galleries and remained just as enthusiastic when confronted with the unexpected revelations which increased as I advanced. I had literally gone mad.

What stunned M. Laval and later observers was the vibrant quality of the painting. Not only were the animals life-like, they seemed to move, be full of energy – even magical. A representation of a bison with one of its hind legs crossed behind the other and a gap left between it and its torso shows a mastery of perspective. This sort of skill was not seen again in Western art until the Italian Renaissance of the fifteenth century. And when Pablo Picasso visited the caves after the end of the Second World War, he was much moved and commented, 'We have learned nothing in 12,000 years.'

Lascaux is not unique. In southern France and northern Spain, essentially on either side of the Pyrenees, there are 350 caves with paintings (and also many engravings carved into the rock with flint chisels) on their walls. They were made by small communities of hunters, all of them *Homo sapiens*, between 30,000 BC and 8,000 BC. This was the time of the last Ice Age and the caves are found only in the areas known as the Refuges. These were the places where refugees found shelter from the bitter winds and harsh winters, and where the climate allowed plant and animal life to thrive sufficiently to support small populations. Many of the painted caves of the Refuges are found in valleys like the Vézére. The cliffs sheltered people from the worst of the weather and the rivers at their foot watered the herds of reindeer and wild horses as they migrated across them, seeking the summer grasslands.

Aside from their patent beauty the point of these paintings is mysterious. Made in darkness, lit only by flickering torches, their style is amazingly consistent over a tremendous span of time, for 22,000 years, the long lifetime of the Ice Age Refuges. Using mineral-based paints such as red and yellow ochre, charcoal and pads of fur or moss or brushes of animal hair to apply fields of colour, they sometimes seem chaotic with animals painted over each other. But in the Hall of the Bulls at Lascaux, where Marcel Ravidat first tumbled into a darkness undisturbed for 17,000 years, the composition looks as though it was organised by one directing mind and clearly intended to be seen from a single viewpoint in the approximate centre of the floor of the cavern. Skills of all kinds were clearly passed on and often paintings were made in places only reachable by scaffolding. The cave art of the Refuges is an extraordinary achievement.

The bulls, the bison, the lions and the mammoths are magnificent, sometimes painted on a monumental scale. By contrast people are usually represented by stick figures, quickly sketched and often incidental to the riotous gallop of the herds and the pounce of the predators. This artistic preference must reflect the painters' sense of the world outside the cave. It was a world overwhelmingly inhabited by animals, thousands grazing the wide grasslands, drinking from the rivers, migrating south in the winter – it was not yet a world made by men and women. There were so few amongst the teeming herds and the ferocious killers who hunted them and who terrified those who saw them attack. Packs of hyenas as big as lions were persistent and deadly, their huge jaws able to snap bones and tear off gouts of flesh. In wet places traversed by winding paths through willow scrub and birch woods, wolverines would climb trees and hide themselves, crouched and waiting. When deer or antelope forced to keep to the path passed below, the wolverine jumped on its back, digging in its razor-sharp claws to avoid being bucked off. As the agonised animal bolted in panic, running away from the appalling pain, the wolverine tore at its neck muscles until the vertebrae could be cracked. The world outside the caves did not belong to men but to the great herds of prey and their savage predators.

In 1994 the Chauvet Cave was discovered. Its paintings had not been seen for 27,000 years and many were of the great predators of the grasslands and the wooded valleys; lions, bears or large and dangerous animals not normally hunted such as rhinos or mammoths. Ten thousand years later when the painters began work at Lascaux, very many fewer predators stalked the walls. Had men replaced them at the top of the food chain? And if the world outside was less dangerous for hunters, why did the prey of the grasslands inhabit their imaginations?

Some animals seem to be featured more strongly than others. Horses gallop through Lascaux, bulls or bison or cattle dominate elsewhere. It may be that the hunting bands adopted an animal as its totem. Linked to the herds by necessity and forced to follow them if they moved or exhausted their ranges, the hunters and their painters may have attempted to anchor them in one place by making their likenesses, something of their essence, in the secret places under their lands.

In unpacking this notion, parallels begin to suggest themselves. Throughout different cultures and periods of history, caves have been seen as the entrances to underworlds, transitional places between the light, open-air world of the surface of the earth and what lies beneath. Just as the much later prehistoric farmers buried their dead in the land they had cultivated in order to underline to outsiders that it belonged to them and in that way make a reality of their possession of it, so the hunters of the Refuges may have tried to bury symbolically the animals in the land they grazed – to keep them close, and at least in one sense, to own them.

On the island of Rousay, in Orkney, there were 13 townships in the nineteenth century and in each there exists the remains of an ancient chambered cairn, a tomb. Some time around 3,000 BC the Farmers buried their dead in the land they toiled over so that there could be no doubt or dispute that this was the kindred ground in every sense.

Animals may also have been adopted by the bands of the Refuges as totems in another way, that is, they identified with them very closely. Perhaps the Horse Clan of Lascaux painted so many of them because they admired their speed and beauty, hoped that they might share their virtues and not only because their livelihood depended on the herds.

The caves are atmospheric places, often cool, often echoic, and away from the mouth, black dark. Painting was not only done by torchlight, it was also revealed by the yellow flicker of flame. Rituals associated with magical revelations in the darkness are not hard to imagine. If sound effects were added – snorts, whinnies, roars and thundering hooves – then there may have been a sense of powerful son et lumière. Bone flutes have been found by archaeologists and with the widespread use of hide in all sorts of domestic applications, the taut and resonant skins of drums must have been an early invention. At festivals and parades through Britain's summer streets the ancient sound of fyfes and drums is still heard.

The form of rituals involving the animal paintings, music, song and dance can be guessed at but no detail has come down to us. But that these rites were long-lived there can be no doubt. As late as 1458 Pope Calixtus III (from Valencia) rebuked nominal Catholics in the northern mountains of Spain for performing rituals in 'the cave with the horse pictures'. Tantalisingly, no details are given.

Trances may have been induced in the airless painted caverns, either by dancing or exertion in places where there was little oxygen. In several chambers at Lascaux pockets of carbon dioxide build up naturally and have caused evacuation in recent times. Transcendent states may have been induced in the Hall of the Bulls and other larger spaces, or perhaps in tiny claustrophobic passages. Another Orcadian example illustrates. At Mine Howe a winding corkscrew of a prehistoric stairwell bores into the ground to lead to a tiny sensory deprivation chamber where individuals may have stood alone in the darkness, hoping for visions in the chill and damp bowels of the earth.

Perhaps the most arresting images at Lascaux and the other caves are not the extraordinary animals. Filling their mouths with paint, men placed their hands flat against the cave wall and, either directly or through a pipe, sprayed them. The effect is like a stencil, leaving a handprint on the wall, a sort of eerie, fleeting signature.

When the excavators entered the amazing painted caves at Chauvet in 1994, they knew for certain that no one else had seen them for 27,000 years. Undisturbed across all those millennia, footprints were found on the clay floor and there were two clay-stained handprints on the cave wall. It seems that a ten-year-old boy, carrying a torch that left charcoal marks as he walked on, had visited the magical cave – alone – just before it was lost to history. The image of an awestruck little boy staring at the mammoths and the lions is powerful and mirrored by the wonder that overwhelmed Marcel Ravidat and his friends almost 30,000 years later.

The cave paintings are a unique record, a pungent, highly coloured legacy left by our oldest direct ancestors. Compared to the dusty collections of flint arrowheads, hand axes and skeletal fragments, they flicker with life and speak of a vivid world inhabited not by prehistoric shadows but by the flesh and blood of men, women, children and animals. On the cave walls are glorious examples of virtuoso artistry, humour, humanity and enigma and they announce the arrival and survival of people like ourselves in Europe.

What is even more remarkable is something very simple – the ability of the paintings to move us profoundly, to be awestruck like Marcel and his friends. It is a reaction reaching across 30 millennia to touch us and draw the little boy with the torch and our ancestors closer. And more, if the painters of the Refuges could make great art to astonish Picasso and all who have seen them, then how can they seem to be lesser beings, primitives or savages?

Their skill and sensitivity are not the only link to bind us. We carry the memory of these remarkable people inside our bodies. When the weather at last warmed and the hunters could leave the Refuges and their caves, they began a journey to the north, a genetic journey to Scotland. Their DNA has survived and some of the descendants of the men and women who hunted and painted the bison and the wild horses now live in Scotland. This is the story of their journey home.

The Book of the Dead

✳

The first explosions were heard on this island in the evening of 5th April, they were noticed in every quarter, and continued at intervals until the following day. The noise was, in the first instance, almost universally attributed to distant cannon; so much so, that a detachment of troops were marched from Djocjocarta [Yogyakarta], in the expectation that a neighbouring post was attacked, and along the coast boats were in two instances dispatched in quest of a supposed ship in distress.

Sir Stamford Raffles

I T WAS NOT GUNFIRE but something much more deadly. In 1815, Sir Stamford Raffles was Governor-General of the new British colony of the Indonesian island of Java and 350 miles to the east, a world-shaking catastrophe was taking place. In the most devastating and powerful volcanic eruption in recorded history, Mount Tambora blew itself apart. Inside the cone, pressure had been building in a huge magma chamber and on 10 April it exploded. The roar was heard 1,600 miles away on Sumatra and Raffles was astounded by what he saw, writing that Tambora became 'a flowing mass of liquid fire'. The top half of the volcano had simply disappeared, shattered into smithereens by the ferocity of the eruption, reducing its height from 4,300 metres to 2,851 metres. At least

11,000 people were reckoned to have been killed immediately. Governor Raffles sent Lt John Phillips to Sumbawa, the site of the ruined volcano and he left these notes:

> On my trip towards the western part of the island, I passed through nearly the whole of Dompo and a considerable part of Bima. The extreme misery to which the inhabitants have been reduced is shocking to behold. There were still on the road side the remains of several corpses, and the marks of where many others had been interred: the villages almost entirely deserted and the houses fallen down, the surviving inhabitants having dispersed in search of food
> . . .
>
> Since the eruption a violent diarrhoea has prevailed in Bima, Dompo, and Sang'ir, which has carried off a great number of people. It is supposed by the natives to have been caused by drinking water which has been impregnated with ashes; and horses have also died, in great numbers, from a similar complaint.

Tambora's effects across the Indonesian archipelago were devastating. Large fragments of pumice rocketed into the air, some measuring 20 centimetres across, and killed people as they fell. An ash plume formed and descended in a radius of 800 miles while a tsunami tore into the surrounding islands. Trees were uprooted on Sumbawa and catapulted into the sea where they mixed with cooling and solidifying pumice to create an extraordinary phenomenon. Vast rafts, some of them three miles across, were driven out into the Indian Ocean by the tsunami and, five months after the eruption, one drifted onshore near Calcutta.

The wreckage and carnage was not confined to south-east Asia. Many thousands of tons of sulphur exploded from Tambora into the upper atmosphere along with huge clouds of superfine volcanic ash. They combined into an aerosol screen and stratospheric winds carried this toxic mixture great distances. It altered the Earth's climate profoundly. In the spring of 1816, a 'dry fog' blanketed the north-eastern United States; in June snow fell in New York State, frosts bit hard and, across a huge area, crops were ruined, left black and rotting in the ground. Across the northern hemisphere, animals died in their millions, famine followed and the extraordinary climate is thought to have been the cause of a typhus epidemic which raged along the Mediterranean littoral and south-eastern Europe. Appalled observers wrote of 1816 as 'the year without a summer'.

Known as a super-colossal eruption, Mount Tambora was more severe than Krakatoa or any other recorded volcanic event in the last two millennia. But its effects were short-lived and mild compared to what happened 70,000 years before on the island of Sumatra 1,800 miles to the north-west. More than 70 miles long and 20 miles wide, the placid tropical waters of Lake Toba now fill the vast crater of what geologists call a super-volcano. The eruption of Toba was two orders of magnitude greater than Tambora and the most severe in the last 25 million years. It changed the world and directly affected our ancestors, their DNA and the long journey it eventually made to Scotland.

Toba took place too long ago for any description to have survived except in the geological record but it appears that the nature of its impact was similar to Tambora albeit on a much greater scale. The supereruption plunged the planet into a six-toten-year volcanic winter. Little or no sunlight could penetrate the dense and persistent sulphuric aerosol and, as the skies darkened over continents, plants withered and died. A blanket of ash 15 centimetres thick covered the whole of the Indian subcontinent. With nothing to eat and water contaminated, animals died in their millions – and populations of people, of *Homo sapiens*, appear to have teetered on the brink of extinction.

Scientists believe that Toba left only a tiny group, perhaps just five or 10,000 people, alive and it seems that they survived in a Refuge in the rift valleys of Central Africa. There may have been as few as 2,000 able to conceive and give birth – and we are all of us the descendants of this tenacious remnant.

Their DNA is our DNA and how it was structured, how it mutated and how it changed over time are the key texts in understanding the genetic journey of the entire human race, in being able to read the Book of the Dead.

In the autumn of 1951, two brilliant scientists began to work together at Cambridge University. Within two years, the collaboration of Francis Crick and James Watson had created a convincing model for the molecular structure of DNA. It was a double helix – two long strands of chemical bases forming a spiral and linked in pairs, like the steps of a spiral staircase. Crick and Watson had made a world-changing discovery; understanding the structure of DNA quickly led to a clear sense of how it worked.

DNA or deoxyribonucleic acid is one of the central building blocks of life, the hereditary material in human beings and almost all other organisms. Even though every cell in the human body has the same DNA and most people's DNA is very similar, it is nevertheless the key to individual identity, to understanding the differences between us and how our history across the planet diverged and coincided.

The information in DNA is stored as a chemical code made up from four bases: A is adenine, C is cytosine, G is guanine and T is thymine. Like the letters making up the words on this page, the code is the order of the chemicals in the DNA strand. When they are written out, these letters make a sequence – for example, ggaacagatttaccacccaagta. Each person carries two copies of the code, one inherited from their mother, the other from their father, and they are joined together in the double helix. In total, there are six billion letters, three billion from each side of the family. About 99.9 per cent of these are the same in all people everywhere and they collectively identify us as human beings. But there are tiny marginal differences or mutations and they are what make us very slightly unlike each other. For instance, rather than carrying the ggaagcatttgggtacagta sequence, another person might have a thymine instead of adenine for the fourth letter, giving ggatgcagatttgggtacagta. This change might arise when the process of reproduction copies all of these millions of letters to make new DNA for the next generation. Depending on where in the DNA sequence a change occurs, it may have no effect since much of it appears not to be critical to our lives as it is endlessly copied into the future. It seems to work like a kind of harmless parasite inside our bodies. But, when a change does occur in a critical place, a devastating disease can be the consequence. All changes, benign and malign, are inherited and they form a chemical archive of our history, from family trees to the origins of humanity.

Studies from DNA populations around the world have revealed how similar all humans are compared to, say, fruit flies or chimpanzees. This is because humans are a very young species and there has been less time for changes to occur. People like us first come on the archaeological record in Africa about 150,000 years ago, a mere moment in evolutionary time. Other species are millions of years old and they have had time to mutate and change over a very much longer period. For example one troop of 55 chimpanzees carries more genetic diversity – more differences in their DNA letters – than the entire human race of seven billion people.

It is important to understand that African DNA is special. It is more diverse than non-African DNA so that, in any sequence of letters, more changes, more variants are found in Africans. The reason for this is straightforward – humans have lived in Africa for much longer than anywhere else and there has been more time for these changes to happen and to spread.

This comes into focus when two crucial elements of our DNA are studied – the Y chromosome and mitochondrial DNA. The Y chromosome is what carries the gene to create men. The default embryonic programme is for all of human reproduction to create only females but the presence of the SRY gene on the Y chromosome initiates a cascade of events leading to a different outcome. Only men carry the Y and it is passed on by fathers to sons and grandsons and so on down the generations. And, crucially, because there is no female Y chromosome, it does not become mixed or diluted in succeeding generations like the rest of our DNA.

From a geneticist's point of view, this is extremely helpful. It means that a 60 million letter block of DNA is inherited through the male line intact, from generation to generation. Many changes can be found in one sequence and therefore many different types of Y chromosomes can be identified. And, moreover, the relationships between these different types can be recovered by looking at the pattern of sharing of the different changes. For example, two Y chromosome types sharing ten variants are obviously more closely related than they are to a third lineage with the same ten variants but an extra one that neither of the first two have.

Mitochondrial DNA (mtDNA) is the counterpart of the Y chromosome and it is passed on by mothers to their children. Sons have it but do not pass it on, their mtDNA dying with them, but daughters always pass it to granddaughters and so on, like men do with the Y chromosome. Mitochondrial DNA does not mix with other DNA and is therefore also inherited intact, preserving the clear relationship between lineages. Seminal work in the 1980s with mtDNA first revealed our recent African origins but, because it has fewer letters than the Y chromosome DNA, the latter can provide much more information about our past.

When the early communities of *Homo sapiens* in Africa were decimated some time around 70,000 BC and the supereruption of Toba, a tiny number of survivors detached themselves from the main remnant groups and began to move north from the rift valleys. What prompted the exodus may have been the devastation of the volcanic winter and the shrivelling of plant and animal life even in the warmth of Africa. In any event, it seems that only a few hundred people took the first steps on an immense journey, one which would ultimately populate the whole of the rest of the world.

When the exodus bands reached the Horn of Africa, they crossed to the Indian Ocean coast of the Arabian Peninsula. The Bab el-Mandeb, the Gate of Tears, that leads into the Red Sea is only ten miles wide between modern Djibouti and the Yemen but, even over that distance, boats will have been needed to gain the farther shore.

Over each new horizon, they carried the secrets of their DNA inside them and, as they crossed rivers and mountain ranges, it seems that only two mother-line lineages and two father-line lineages survived the privations of their great journey. All of the human beings who are not Africans or are not descended from Africans are their children. These pioneer bands left behind a much more diverse group. African DNA has about 20 mtDNA lineages and approximately ten more Y lineages which can be traced back to the time of the exodus.

Physical characteristics show African diversity very clearly. While Europeans believe themselves to look very different from Chinese,

Korean or Japanese people, for example, Africans are much more different from each other. In the south, the San peoples, the bushmen of the Kalahari, are small hunter-gatherers who were displaced by the Zulu and Xhosa from the north, groups who looked strikingly different – taller, rangier and with a very different culture. The pygmies of the jungles of Central Africa are perhaps the most extreme example of a variant.

As the pioneers reached the Persian Gulf, some appear to have swung northwards to the lands watered by the rivers Tigris and Euphrates. The region that used to be known as Mesopotamia was the place from where human beings, *Homo sapiens*, eventually began to move into Europe and mid-Asia. DNA shows both the pace of this migration and has something to say about the people who began to walk to the west – and a remarkable encounter.

Recent research has revealed startling new information about human DNA. When the bands of pioneers from Africa reached Mesopotamia and the Levant around 60,000 BC, they encountered groups of Neanderthals – and they mated with them. Scientists at the Max Planck Institute in Leipzig have sequenced the whole Neanderthal genome from the powdered bone fragments of three females who lived in Europe around 40,000 BC. They then compared their DNA with the genomes of five people from France, China, western Africa, southern Africa and Papua New Guinea. There was no correlation with the two sets of African DNA but, sensationally, it became clear that between 1 per cent and 4 per cent of the DNA of the non-African lineages comes from the Neanderthals. The wide geographical spread of the non-African samples strongly suggests that *Homo sapiens* mated with Neanderthals in Mesopotamia before the dispersal along the Indian Ocean coast to Australasia as well as to Europe and eastern Asia.

The analysis gave no information on whether or not Neanderthals living in Europe 20,000 years later interbred with our European ancestors but there is currently no evidence of this from mtDNA, Y chromosome or other kinds of DNA. Nevertheless, these new findings do mean that most Scots will have inherited a small but variable proportion of their genes from these ancestral Neanderthals of the Near East.

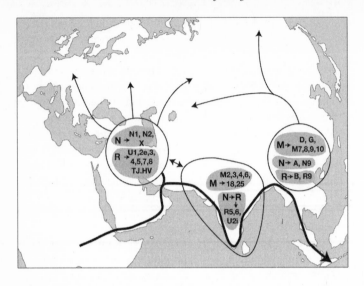

The peopling of Eurasia as inferred from mtDNA variation. The bold black arrow indicates the rapid southern coastal migration from the Horn of Africa to Australasia. After initial colonisation three expansion zones are highlighted where the mtDNA types indicated arose, and from where they later spread to the interiors of the continent (smaller arrows). The ancestors of the Scots moved from SW Asia to Europe across the Bosphorus at various times in prehistory. Although not shown here, some rarer groups, such as Y chromosome M35 left Africa later via a Northern, Levantine route.

It is clear that those men and women who hunted the megafauna before the last Ice Age, who attacked and brought down their prey, who may have used their teeth as savage weapons have played a role in our common human ancestry. Perhaps we should be careful when we mock them. They were brave and hardy – qualities most would hope to inherit.

Between c.39,000 BC and c.37,000 BC, Neanderthals became extinct, a last remnant clinging on around the Rock of Gibraltar, in the south-facing Gorham's Cave. Artefacts used by these fading communities were found there, and they appear to have been used by a small group, probably one of the last to survive. Our ancestors,

Homo sapiens, arrived in Europe in the millennia between c.43,000 BC and c.41,000 BC and so co-existed for a time with Neanderthal hunter-gatherer bands, perhaps for as long as 5,000 years. It may be that there was violent competition for the right to use the extensive ranges needed by communities dependent on the wild harvest of fruits, roots, nuts and berries, and on the ability to hunt prehistoric fauna. Our ancestors may have driven Neanderthals to marginal lands where the slow process of extinction began as bands shrivelled and starved. One anthropologist believes that *Homo sapiens'* domestication of dogs was important in this struggle for dominance. Along with Neanderthals and modern humans, wolves were top predators and our ancestors gained a crucial competitive edge when they began to train wolves to hunt and also to protect downed prey from circling rival carnivores such as lions and hyenas.

Less dramatic but perhaps more persuasive is recent research on disease and its transmission in prehistory. In 2016, British scientists argued that when *Homo sapiens* emerged from Africa, they brought deadly diseases, having themselves developed a tolerance over many millennia. This hypothesis arose out of research that concluded that some infectious diseases are far older than had been previously believed. Helicobacter pylori, a bacterium that causes stomach ulcers, is very likely to have been passed on to Neanderthals by modern humans when they entered Europe and came into contact in c.43,000 BC. Herpes simplex 2 is the virus that transmits genital herpes, and it too is highly infectious. It is a sexually transmitted disease and even only a few contacts between *Homo sapiens* and Neanderthals could have spread it widely.

When the tiny numbers of Spanish conquistadores overthrew the mighty Aztec and Inca empires, their strongest weapon was not the sword but deadly European diseases passed on to Native Americans who had no tolerance. They died in droves, especially in the great cities of Mexico and Peru. It may well be that a very similar process took place when our ancestors first encountered Neanderthal bands in prehistoric Europe, although the sparse and dispersed nature of hunter-gatherer society will have meant a much slower spread of any disease.

DNA can trace the footprints of the dispersal out of Africa as people with a particular and identifiable lineage found good places to live. Some decided to stay in a location like Mesopotamia, for example, where they could hunt and gather food while others eventually moved on.

Movement along the Indian Ocean coast was very rapid and within only 2,000 years pioneer bands had reached all the way to New Guinea and Australia. Because of this, there are very old lineages in south-east and south Asia.

By contrast, there appears to have been a pause, perhaps as long as 20,000 years, before bands began to move out of the Near East towards Europe. Perhaps the way to the west and north was blocked by wide swathes of desert.

As the footprints of *Homo sapiens* stamped themselves on the map of the world, their DNA maintained a link with the places their ancestors had been before them. These connections can be seen very clearly when comparisons are made. Each region has a set of predominant lineages which might be said to be characteristic of that place and some of these are clearly very old – even original.

The statistics of historical genetics are very straightforward in this regard. The origins of a particular lineage or marker can be ascertained when two related factors are taken into account. First, geneticists look at distribution. Where in the world is the marker most common? Where do most of its carriers live? Second, the number of mutations amongst these populations is counted. If there are more than in any other place where the marker is found, then that means it has been in that place for the longest time.

When such lineages turn up elsewhere, it is possible to see in the example of a single person how far the exodus out of Africa reached and the various routes it took. There is a farmer on the Hebridean island of Islay who was astounded to be told that his DNA was linked in a direct line with an ancient Y lineage in Mesopotamia, modern Iraq. In the genes of the farmer and his sons, the story of an immense journey still lives.

When a scientist analyses a sequence of DNA letters, A, C, G and T, in any two individuals, they will see that some are identical but that others have differences, perhaps 14 letters out of the first 700. These arise from errors in genetic copying – that is, in the production of sperm and eggs. With three billion letters from the mother and three billion more from the father to replicate in each generation, it is inevitable that such incidental copying errors will occur. And it is just as well that they do, for these differences are the raw material for understanding evolution and the DNA history of the human race.

Using what is known as the 'molecular clock', geneticists can measure the pace at which changes in DNA sequences of letters occur. Over long periods, mutations appear to take place at regular intervals and this, in turn, allows approximate dates to be attached. When the occurrence of these mutations is calibrated against other data such as archaeological finds or fossil records, the chronology becomes more secure. For example, the chimp–human divergence occurred about six and a half million years ago and so provides a calibration point for the rate of change. This process underpins any clear understanding of the dating of genetic history.

One of the most ancient Y lineages in Scotland is known as M284 and it accounts for about 4 per cent of all Scottish men, a group of around 100,000. And it is a living link with the cave painters of the Ice Age Refuges. When the weather warmed after 12,000 BC, some of these men walked north and crossed dry-shod into the European peninsula that was Britain and they carried the marker M284 with them. It developed a later subset called S165 which is essentially only found in the British Isles and which charts the progress of these pioneers. S165⁻ is the older subset and it is more widespread and more common in England while the younger S165+ is more common in Scotland.

Examples of this marker are also seen in Ulster, not only amongst plantation families (migrants from Scotland and elsewhere, most of whom arrived in the seventeenth century) but also in those of older pedigree. This shows an ancient connection across the North Channel. It is very likely that S165 arose somewhere in the British Isles and shows a degree of commonality between some Englishmen and some Scotsmen, probably a memory of an ancient shared British ancestry.

M284 is very rare outside the British Isles and Ireland but it is found in tiny numbers in France and Germany. Analyses of Portuguese samples have revealed a significant number of M284 chromosomes at higher frequencies than anywhere else on the continent. This points to a possible Iberian origin, perhaps groups leaving the Ice Age Refuges and moving westwards and south.

Mitochondrial groups H5, U5b1, H1 and V all appear to have originated in south-west Europe and to have moved after the end of the Ice Age between 11,000 and 13,000 years ago. Each has both its highest frequency and diversity in this part of Europe and it become fainter as traces radiate outwards. Both the Y and mt groups account for a significant proportion of Scottish lineages and they have deeper origins in the Near East.

What this analysis shows is something simple and unarguable. Some of the earliest Scots were the direct descendants of some of the earliest people to reoccupy England after the last Ice Age and both have close links to the people of the western Refuges.

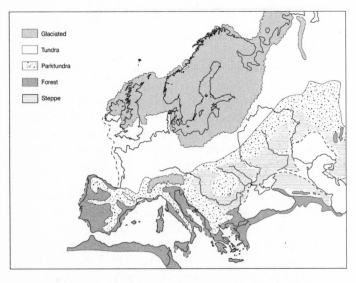

18,000 years ago – the peak of the last ice age. Scotland and the north of Europe were covered in thick ice, with tundra to the south. Park tundra is a slightly less extreme form of tundra, with low grasslands and alpine flora. Humans could only live in the southern parts of the continent.

When the thaw came, it was rapid. And as the summers warmed and lengthened, grass grew, herds of grazing animals followed it northwards and human hunters came closely and quickly behind. New analyses of carbon-dated remains in Britain and updated readings of the evidence from Greenland ice cores confirm that climate change took place over decades rather than centuries and that within the span of the memories of one or two generations pioneers left the Refuges and moved very rapidly into northern France and southern England. This was no slow exodus and the migrating herds probably set a brisk pace, pulling a dramatic repopulation behind them. As the ice and tundra retreated, bands of hunters advanced.

At Creswell Crags in Derbyshire, engravings on the walls of the limestone caves have been recognised and they show a clear link to Lascaux and the other famous sites on either side of the Pyrenees. High up on the roof of Church Hole, the outline of a bison has been made out and also a large stag. There are thought to be faint traces of 20 or 30 figures at the Creswell caves but perhaps the most intriguing is that of an ibex. Believed to be extinct in Britain after the end of the last Ice Age, this species of wild goat was still browsing the forests of the Ardennes on the Franco-Belgian frontier. It may be that the engraver at Creswell Crags had seen an ibex and was working from memory.

3

Northwards

AS THE HURRICANES whipped around the dazzling summit of the ice dome over Ben Lomond and a glaring sun beat down from a cloudless blue sky, Scotland lay sleeping. Under the crushing blanket of the ice, the landscape was compressed, hidden, waiting – except in one place. From bore-holes drilled through the crust of primeval soils, sediments, clays and rocks, geologists have found evidence that the western edge of the ice sheet over northern Europe reached only as far as the Hebridean island of Islay. The grip of the glaciers slackened on a line along the sea lochs of Gruinart and Indaal. The Rinns, the eastern wing, was cut off from the rest of the island around 20,000 BC, the time of the Ice Age's zenith and when temperatures were at their coldest.

The storms and dark clouds of the depressions at the edge of the glaciers will often have obscured it but the hill dominating the Rinns was high enough to escape the ice. Beinn Tart a'Mhill appeared like a dark beacon in the pitiless white landscape and, although they too will have been scarfed in cloud, the Paps of Jura to the east will have risen above the ice desert. This pocket of land, the Rinns, was not bulldozed and scarted by the rumble of glaciers, and deposits of flint-rich sediments were left undisturbed, an anomaly which would later become important.

When the ice began to shrink back north-and eastwards, around 15,000 BC, Islay slowly revealed itself. By 12,000 BC, the southern Hebrides were clear, a tundra emerged and, as the great weight of the ice lifted, the land rose and the Rinns were reunited with eastern Islay.

Near the village of Bridgend at the head of Loch Indaal and almost precisely on the line of the edge of the glaciers, a flint arrowhead was found. It was not just any arrowhead. In the summer of 1993, a group of archaeologists was field walking, looking for objects turned up by the plough. The arrowhead was of a specific and very early type known as an Ahrensburgian point. In Schleswig-Holstein, the neck of land linking Germany and Denmark, archaeologists discovered a site of mass slaughter. Through the narrow Ahrensburg Valley, huge herds of reindeer had once migrated north to summer pastures, crossing a prehistoric land bridge to what is now southern Sweden. At Stellmoor thousands of reindeer bones were found at the edge of a lake. In the anaerobic mud, excavators came across well-preserved pine arrow shafts and a huge scatter of flint tools. Often lodged in the skeletal remains of the reindeer were the distinctive arrowheads that became known as Ahrensburgian points. Not only did the finds signal the arrival of the bow and arrow in northern Europe, they also allowed the reindeer ambushes to be securely dated. It appears that the hunters lay in wait around the year 10,800 BC.

The Ahrensburgian point picked up in a ploughed field on Islay is not unique; others have been found on Tiree, Jura and Orkney but their find-spots were not securely recorded and some were damaged. Nevertheless the arrowhead recognised in 1993 is very suggestive. Some time in the eleventh millennium BC, soon after the retreat of the ice, hunters were venturing north into the emptiness that was Scotland. Probably paddling the skin boats known as curraghs, at least one summer expedition appears to have made landfall on Islay and perhaps Tiree and Jura. Pioneers explored the islands and the coastline and it seems that they brought bows and arrows with them. Even if they stayed for only a short time, these were likely to have been the first people to see Scotland for 15,000 years or more.

Summer temperatures were rising to a maximum of 17 degrees Celsius but the winters remained very cold at an average of only 1 degree Celsius. Much of Scotland became tundra after its release from the ice, and then the sort of summer grasslands grew that sustained the vast herds of wild horse and bison so beloved of the cave-painters of Lascaux and elsewhere. It may have been cold enough for reindeer when the pioneers came to Islay in 10,800 BC but their quarry is more likely to have included deer. Jura is an anglicisation of Diura and from Gaelic it translates as Deer Island. The hunters probably stalked other species such as wild horse, wild cattle and perhaps boar. Seals may have colonised the shoreline and, being slow moving on land, they will have made easy prey.

For millennia water was the most reliable highway, a connector rather than a barrier. And, if a trackless wildwood was beginning to creep northwards by the time the hunters came, then coastal waters, lochs and rivers will have been an easier and faster means of travel. Hide boats were easy to make and could be assembled in a morning. A frame of whippy green rods is lashed together with cords (animal sinew is reliable) into the shape of a hull and then stiffened with wooden thwarts which can acts as benches for paddlers or rowers. Then a patchwork of fresh hides is sewn together and stretched over the frame. When this membrane dries, the hides become taut and pull the curragh into shape. Caulking with resin or animal fat along the seams works well enough although it is likely that baling was a constant duty. These ancient craft were made (and continue to be made in south-western Ireland) from material readily to hand and they were very versatile. Compared with much heavier wooden boats, they have a shallow draught measured in centimetres and, if a river passage becomes difficult because of rapids or rocks, a curragh can easily be picked up and carried over long portages. If the pioneers who dropped the arrowhead at Bridgend wanted to travel from the head of Loch Indaal to Loch Gruinart, a distance of less than two miles over flat terrain, they will simply have picked up their boat and carried it. If it rained on a voyage, the sailors could have put ashore and upended their boat to make an excellent shelter.

In 10,800 BC, hunters on an expedition, particularly one into the unknown, will have been properly equipped and dressed. And they would have been no rabble of ragged savages wrapped up in furry pelts roughly tied around the middle. All of the evidence suggests that prehistoric hunters wore and carried sophisticated gear adapted to the conditions and which they could maintain and repair as they moved through the landscape. Archaeologists have found needles at sites in the Refuges in southern France and these date to around 17,000 BC. Even earlier awls have been discovered and these sharp-pointed instruments, often made from bone, were used for making holes in hides so that they could be shaped and laced together. If hunters needed to move quickly or quietly through the wildwood in pursuit of prey, they will have worn close-fitting and well-made clothes which would not have hampered movement.

In a glacier high in the Alps near the border between Italy and Austria, a mummy was found almost perfectly preserved in the ice. In September 1991, two German tourists from Nuremberg, Erika and Helmut Simon, at first thought they had found a modern corpse. Eventually nicknamed Ötzi, the body was, in fact, dated to around 3,300 BC and, amazingly, much of the man's clothing and equipment had survived for more than 5,000 years. He had been killed in a violent manner, probably by a blow to the head, and his forearm was freeze-framed in a protective position across his neck and face. Very soon after he was left for dead, the ice buried him.

Ötzi will have looked much like the hunters of the previous seven millennia. The only item found on his body not available to the pioneers who came to Islay in 10,800 BC was a copper axe head hafted to a yew handle. To keep him dry, Ötzi wore a cloak made from woven grass. This was not the stuff munched by cows and sheep but tough marsh grass which grows in spiky clumps in wet places. The design of the cloak is ingenious. The thick green stalks were tied in small bunches and laid over each other in a regular overlapping pattern like sheaves of thatch on a cottage roof. The cloak would have been light, waterproof, easily packed away and equally easily repaired. Ötzi may also have used it as a shelter or a sleeping mat.

Some evidence of aesthetics was found. Ötzi's coat was designed and made as much for looks as utility. Reaching down to his knees and with long sleeves, it was sewn together with the fur side out and in strips of differently coloured pelts deliberately alternated light and dark. The coat was tightly belted because it was worn next to the skin and not over any garment resembling a shirt or an under-tunic. Fabric weaving waited in the future. On his lower body, Ötzi wore a leather loincloth and buckskin leggings up to his thighs. And on his head sat a thick bearskin cap with leather chinstraps to keep it secure in the winter winds of the high Alps.

To allow him to move safely and reasonably speedily though the snowdrifts, Ötzi may have been wearing snowshoes at the time of his death. Beautifully made (perhaps by a specialist cobbler, according to one scholar) from different skins, his shoes were waterproof and had wide soles made from thick bearskin. Softer deer hide was used to form the uppers and inside was a netting of tree bark to fit the wide shoes more closely to his feet and, to keep them warm, soft grasses were packed in, like organic socks. Near Ötzi's body, part of a wooden frame and some hide were found and it has been convincingly argued that these were the remains of a snowshoe.

In a pouch sewn with sinew to his belt, the ancient huntsman carried a sharp bone awl, a flint scraper, a drill, flint flakes and a dried fungus. In addition to his copper axe Ötzi had a flint knife attached to an ash handle, a quiver of 14 arrows, a bow string, an antler tool for making arrow points and a yew longbow of almost two metres. Most intriguing was what may have been a prehistoric first aid kit. Threaded on to a leather thong were two species of dried fungus and one of these, the birch bark fungus, is known to have antibacterial properties. The other was used for starting a fire. Along with other dried mosses and plants, it would take a spark struck from a flint against pyrite.

This most remarkable discovery opens a window on the life and style of prehistoric hunter-gatherers. The overall impression is of a gritty self-sufficiency. Not only could Ötzi repair his clothes, make flint arrowheads and start a fire, he could also treat an injury.

When forensic pathologists examined the mummified body, they were able to add even more to the story of the physical remains. Ötzi had stood 5 ft 5 in. tall and was around 45 years old. His bones suggested an arduous life of walking over hilly terrain. He may have been a shepherd as well as a huntsman. In his alimentary canal were two partly digested meals of chamois and red deer meat, eaten along with some wild roots and fruits and some cultivated wheat and barley. The latter would not have been available to the Islay pioneers and may well be the residue of bread.

When Ötzi's mtDNA was sequenced, there were surprises. Isotope analysis had already concluded that he was born and raised in the Val Venosta in the Ötztal Alps, near where he died. Six modern inhabitants of the area were found to have inherited his mtDNA or at least the mtDNA lineage of his mother, given the inheritance pattern through the female line, a particular K1 type. But four other descendants show how very far travelled his lineage was, despite its extreme rarity. Matches turned up in Greece, Russia, Norway – and Scotland.

Pollen grains suggested that Ötzi had met his sudden death in the spring. An arrowhead was lodged in his left shoulder and there was a matching entry tear in his coat. Researchers also found evidence of cuts and bruises to his hands, arms and chest as well as the fatal blow to his head. It seems that, on the snowbound Alpine pass, Ötzi was brought down by an arrow and then bludgeoned to death by his pursuers.

No such archaeological drama was left by the hunters who came to the Rinns of Islay in 10,800 BC. In fact, the survival of any trace of their visit is close to miraculous for drama of another sort was waiting. Around 9,400 BC Scotland began once more to grow cold.

Thousands of miles to the west, in northern Canada, a disaster was about to engulf all who stood in its path. As the ice sheets slowly retreated over the North American landmass, a vast lake of intensely cold freshwater built up over much of what is now Canada. It covered an area much larger than the North Sea. At first, it drained slowly over a rocky ridge to the south, shaping the Mississippi river system, flowing ultimately into the Gulf of Mexico. But, as temperatures continued to rise, the ice dams to the west of the huge lake suddenly collapsed. With a tremendous roar, it broke through into the Mackenzie river system of northern Canada and, in a matter of only 36 hours, millions of cubic kilometres of cold freshwater gushed out into the Arctic and from there into the North Atlantic. The effect was catastrophic.

Almost instantaneously the Earth's sea levels rose by a staggering three metres. Tsunamis surged towards helpless and unsuspecting coastlines and countless communities were swept into oblivion. But even more devastating was the dilution of the ocean with an enormous cubic tonnage of very cold freshwater. The Atlantic conveyor belt of ocean circulation, which carried the warming waters of the Gulf Stream northwards, was stopped dead. The climate suddenly changed. Over Scotland and northern Europe storms blew once more, snow fell steadily and did not melt with the coming of summer. A vast ice dome formed over Ben Lomond, the bitter cold crept back over the land and the pioneer bands fled south.

For more than a thousand years the Drumalban Mountains lay under the crushing weight of a dense ice sheet and the rest of Scotland withered into an Arctic tundra. Freezing winds and nine-month winters made settlement impossible but a brief summer melt might have allowed the growth of grazing and encouraged seasonal animal migration and perhaps hunting expeditions. Over 50 generations, stories and knowledge of the buried lands of the north might have been told and retold in the circle of firelight. Perhaps Scotland was not entirely lost in the endless snows and storms.

Eventually, the waters of the great Canadian lake were dispersed, the Atlantic Ocean regained its salinity and the conveyor system brought the Gulf Stream back to northern latitudes. Very quickly, over perhaps a decade, temperatures rose once more and the summer hunters gradually became pioneer settlers. At last, Scotland had finally emerged from the ice.

Very early evidence of the arrival of pioneer bands of hunter-gatherer-fishers was found by chance at Alaterva, the Roman fort at Caer Amon, the early names of the well-set Edinburgh suburb of Cramond. One weekend in the summer of 1999, a team of amateur archaeologists were digging an exploratory trench near the site of the bathhouse used by the soldiers stationed at Alaterva. Instead of an altar inscribed in Latin or some discarded military gear, they came across material that turned out to be almost 9,000 years older. At the bottom of the trench they carefully scraped away the earth around a deposit of flint tools and debris and a few hazelnut shells. For nine millennia these objects had lain unseen and untouched and it was the shells that allowed them to be securely dated. Hazel trees produce only one crop of nuts per year and this organic matter can be carbon dated. And it was reckoned that the six shells analysed had been cracked open some time between 8,600 BC and 8,200 BC. These were the earliest traces yet found of the first people to come to Scotland after the ice.

They chose the site of their camp well. Pitched on high ground near the outfall of the little River Almond into the Firth of Forth, it was surrounded by enough reliable resources to sustain year-round living. In the summer, the hinterland, now covered by Edinburgh's western suburbs, would have made a good hunting ground. In the wildwood and its clearings, deer browsed and the giant cattle known as the aurochs thrashed through the undergrowth. More dangerous were boar and bears but there were many smaller animals to be trapped and birds could be netted. And summer roots, fruits, berries and nuts ripened in the late summer and autumn. The Cramond pioneers would have come to know exactly where the most heavily laden trees and bushes were and when each was in season. Hazelnuts were particularly prized because they could be roasted, mashed into a nutritious paste and preserved through the hungry months of the winter. The wildwood was also the provider of perhaps the most valuable resource – firewood.

The River Almond supplied freshwater fish and, since it is tidal in its lower reaches, the pioneers at Cramond may have built fish traps. Known as *yairs* in Scots, they were woven out of willow withies and staked in a line across the river. Having been careful to make the gauge of the weave wide enough to allow a free flow of water but too narrow for a fish to swim through, the fishermen waited for high tide. Coming in from the Forth to feed in the river, fish could swim over the top of the yairs. But when the tide ebbed and the level of the river dropped, they were often caught behind the mesh of withies. The bone tips of leisters, or fish spears, have been found at several early prehistoric sites and they would have been put to good use at the mouth of the Almond.

At low tide, there were other good things to be had near at hand. Along the unusually wide and gently shelving foreshore that reached out and beyond Cramond Island, there were rich oyster and mussel beds. These were still being harvested in the eighteenth century. In 8,600 BC, this resource in particular was vital because it provided fresh, protein-rich food throughout the winter.

Up on the bluff above the river mouth, where the Roman engineers would mark out the walls and streets of their fortress in the second century AD, archaeologists found faint traces of waste pits, scoops and stake holes dug by the pioneers some time around 8,500 BC. It appears that at Cramond they lived in structures resembling bender tents. Using principles similar to those of curragh building, whippy green rods were driven into the soil and tied with cords before a hide membrane was pulled over and made taut. Perhaps fat and resin were smeared over to make the shelter waterproof and turf laid over it for extra insulation against the winter winds. Fur pelts laid over cut brackens would have made a snug enough place to sleep.

No bones or burials or any physical remains of these early Scots have survived but DNA, archaeology and logic leave little doubt as to where they came from.

We are all, of course, descended from southerners, even those who live in Shetland, Iceland, Lapland and the very farthest extent of human settlement in the north. The ice and the direction of its retreat make that assumption unarguable. But DNA supplies more detail, more colour and unexpected connections. In addition to the M284 marker, another lineage found its way to Scotland from the Ice Age Refuges and the painted caves in southern France and northern Spain. M26 accounts for only 12,000 or so Scots men, around 0.5 per cent of the total population, and it is one of the oldest lineages still to be found. Other markers no doubt died out but it is likely that a substantial group of modern Scots are the direct descendants of the first pioneer bands who came north after the ice. Perhaps those who fished and hunted near Cramond had the M26 marker and, if so, then some of their descendants have not moved far. There is a scattering of addresses across the central belt of Scotland amongst the modern sample.

The M26 marker also moved eastwards from the Ice Age Refuges and it is the founding lineage of Sardinia and carried by an astonishing 40 per cent of the male inhabitants of the Barbagia, the mountainous central region of the island. There is a very tenuous architectural link with Scotland. On the rocky outcrops and ridges of the Barbagia stand the enigmatic drystane towers known as the Nuraghe. There are 8,000 of them and they closely resemble the famous brochs of northern Scotland. The difficulty with the connection is chronological – the Nuraghe were built from a very early date, 7,000 BC, and the brochs much later, in the first century BC and the first century AD – but it is not impossible. Historians believe that the particular skills needed to construct brochs and the speed with which so many appeared over a short period in Scotland may mean that specialist crews hired themselves out to powerful men who wanted such an obvious status symbol in the landscape. Perhaps the broch builders brought their technology and their DNA with them from Sardinia. In any event Scots with the M26 marker are more closely related to the men of the Barbagia in the male line than they are to other Scots.

Mitochondrial DNA is much more difficult to track than Y lineages. Its various types are much more evenly widespread in Europe and it appears that, while men tended to live and die where they were born (until very recently), women moved around a great deal more. This has caused a highly complex mix and wide spread of mother lines in contrast to a fairly straightforward stranding of father lines.

There is a brutal history lesson behind these DNA patterns. The enormous modern improvement in the status of women in Western society sometimes causes us to forget that, throughout the vast sweep of our history, women had little or no status at all. Leaving aside a few famous, mostly royal, exceptions, daughters, wives and sisters were treated like human property. Men almost always inherited wealth and land, which encouraged them to stay put, while women had to move away when and if marriage partners could be found. And women were also traded as virtual or actual slaves. In 697, at the Synod of Birr in central Ireland, the great Dark Ages churchman, St Adomnán, promulgated his Law of the Innocents. Amongst other measures it proposed protection for women who had long suffered as *cumalaich*, 'little slaves'.

Probably the oldest Y lineage in Scotland leads the eye eastwards rather than due south to the caves of France and Spain. M423 has a similar frequency to M26 with around 20,000 Scots men carrying the marker and it is another founding lineage. Remarkably it is shared by between 30 per cent and 40 per cent of Croatian and Bosnian men and appears to have originated in the Danube basin. This was one of the points of entry into Europe for the pioneering bands who left the rift valleys of Africa and came north through Mesopotamia. A subgroup of the M423 lineage appears at the western end of the North German Plain and then reappears on the British shores of the North Sea. It is as though there is a missing historical step.

In early 2001, researchers at Birmingham University looked at a dim, grainy image and were amazed at what they could make out. Picked out in green and yellow against a black background was the course of an ancient and unknown river. Fed by a network of tributaries, it seemed to run for about 30 miles. The disappeared river had been found by seismic reflection, a version of the geophysical surveys done by archaeologists on sites where the remains of buildings or earthworks are thought to be buried. But the green images were different. Supplied by the oil and gas industry, they were part of a vast survey of the bed of the North Sea and what lay under it. The ancient river had once flowed through the hills and valleys of a lost world, a huge part of Europe which had disappeared beneath the waves and been entirely forgotten.

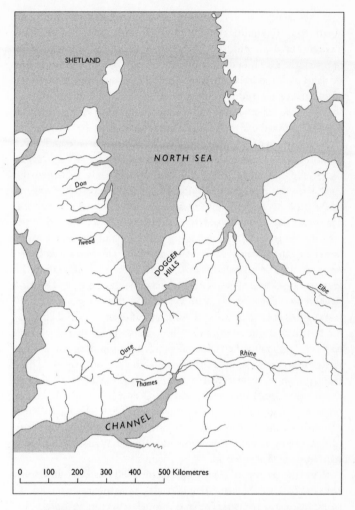

*Doggerland is the name given to the lost continent now covered by the
North Sea and which linked the British Isles with continental Europe.
The map shows Doggerland in the earlier Holocene period, with extensive
coast and estuaries.*

At the zenith of the last Ice Age when bitter hurricanes tore down from the summits of the spherical domes, northern Europe was crushed. Under the great weight of cubic kilometres of ice, the crust of the Earth had been depressed so much in the north that the land to the south had risen. Like a fat man sitting on one end of a bolster, the ice had produced an effect called a 'forebulge'. And so, when the ice retreated, a vast subcontinent was revealed. It was Europe's lost subcontinent, an Atlantis in the east.

To the north of the Shetland Isles, there was dry land and it was only separated from Norway by a deep undersea valley known as the Storegga Trench. In the centuries immediately following the end of the Ice Age, the subcontinent covered almost all of what is now the North Sea and there was no English Channel in the south. The great rivers, the Tweed, Tyne, Elbe and Weser, flowed north into the Storrega Trench while the Thames and the Rhine combined into the Channel River which made its way westwards to the Atlantic. What the Birmingham researchers could make out in the greens and yellows of the seismic reflection image was a prehistoric river which ran eastwards, near what is now the Dogger Bank.

Long before the oil and gas companies began to map the bed of the North Sea, those who lived on its coasts knew that some land had been submerged. At low tide, blackened and clearly ancient tree trunks would appear, often stretching far out to sea. These were known as Noah's Woods and were seen as certain evidence for the great biblical flood. Oyster dredgers had brought up animal bones and clods of a peaty soil called moorlog. Archaeologists had conjectured the existence of a landmass rather than a mere land bridge linking Britain to Europe and, after the famous sandbank, they named it Doggerland.

What the seismic reflection survey supplied was not only confirmation but detail. Much of Doggerland was an immense and watery plain. Many lost rivers meandered through it creating oxbows, wide deltas and large areas of wetland. Some of these flowed into a huge inland sea, what the undersea surveyors called the Outer Silver Pit. The Dogger Hills were a rare area of high ground. In the samples of moorlog brought to the surface for analysis, pollen traces have confirmed birch, willow, hazel, oak and chestnut trees. It is thought that deer, boar, bears, beavers and many other animals browsed the lush vegetation of the Doggerland woods while its wetland will have been home to many species of birds. For hunter-gatherer-fishers it was a very good place to live. But only the faintest, most shadowy traces of people have been found.

On a September night in 1931, Captain Pilgrim Lockwood guided his fishing trawler, the *Colinda*, out of Lowestoft Harbour. Twenty-five miles off the Norfolk coast, by the Leman and Ower Banks, Lockwood had his crew haul in the nets. Amongst the silver glint of fish was a dark lump of moorlog. The skipper pulled it on to the deck so that he could break it up with a shovel and heave the bits over the side.

> We were halfway between the two north buoys in mid-channel between the Leman and Ower . . . I heard the shovel strike something. I thought it was steel. I bent down and took it below. It lay in the middle of the block which was about 4 feet square and 3 feet deep. I wiped it clean and saw an object quite black.

It was a spear point. More than 21 centimetres long and made not from metal but antler bone, this beautiful object had been skilfully worked into a sharp point with one edge precisely serrated so that it would lodge securely in the flesh of its target. One end was scored with transverse lines to enable a good fixture to a wooden shaft. Pilgrim Lockwood had clashed his shovel on what was almost certainly the tip of a leister, a fish spear. Everything about this fine object spoke of people, of manual skill, of food gathering, of someone hunched over the antler point fashioning it and lashing it to as straight a cut branch of ash or oak as they could find in the wildwoods of Doggerland.

More extraordinary was the find-spot. The *Colinda* had been trawling 25 miles off the Norfolk coast and the bone point had been buried in a large lump of what sailors call 'moorlog' (submarine peat), making it unlikely that the leister had been dropped over the side of a boat. Carbon dating placed it deep in prehistory. Lockwood had picked up something made by a hunter-fisherman in Doggerland some time around 11,700 BC.

For a generation what became misleadingly known as 'the *Colinda* harpoon' was all that had survived of the peoples of the lost subcontinent. But more physical evidence has gradually come to light. In the 1970s, a Dutch archaeologist, Louwe Kooijmans, wrote up reports of a series of bone tools which had been brought up from the bed of the North Sea. This time the find-spots were to the south of the Leman and Ower, at the Brown Banks, where the sea narrows between the Belgian and Suffolk coasts. Divers from Newcastle University have recently found more remnants of the Doggerland people off the Northumberland coast, near the mouth of the Tyne.

From the time that the ice in the north began to retreat, around the period when the hunter-fisherman was carefully carving the serrated notches on his leister point, Doggerland was doomed to drown. As the great weight of the huge ice dome over Norway and Sweden diminished and eventually disappeared, the newly revealed land began to rise and what was to become the North Sea began to sink. It was a gradual process perhaps punctuated by occasional local dramas as the sea broke through and made tidal islands of higher ground, creating salt marshes, sometimes making freshwater lakes brackish. For five millennia, the hunters of Doggerland trapped, netted and speared the rich fauna and gathered the abundant fruits, roots and nuts of the fertile plains. But it was to come to end more emphatically than anyone imagined. Around 6,000 BC, geology forced history to accelerate and, in an instant, the world changed.

In the freezing darkness of the Storegga Trench, 2,000 feet below the surface of the North Sea, the bottom feeders sensed an unknowable danger. Above them sharks, whales and wolf fish instinctively swam upwards when the seabed began to quiver and stray rocks suddenly tumbled through the wrack. Dust darkened the water even more as small marine creatures panicked and scattered. And then the undersea world disintegrated. Seismic movement in the Earth's crust caused 3,400 cubic kilometres of rock, gravel and sand to slip into the deeps of the great trench. Along a 290 km length of coastal shelf, an area the size of Scotland broke off and collapsed. The momentary gap sucked in vast quantities of seawater and set in train a rapid and devastating chain of events.

Around the coasts of eastern Scotland, Norway, Denmark and northern Doggerland, the sea retreated rapidly, revealing huge tracts of the seabed. At first, there must have been astonishment on the faces of all who saw it. Perhaps the world was ending, perhaps the gods were angry. Then seabirds suddenly shrieked and took to the air. A distant rumble was heard and, before many had the presence of mind to run for their lives, a gigantic, eight-metre-high tsunami roared on to the shore. Travelling at an incredible 600 miles an hour, it crashed over coastal communities, snapped trees like matchsticks and rained down boulders, debris and sand far inland.

No one saw this undersea earthquake begin and evidence was only discovered by accident when archaeologists exposed a mysterious layer of white, deep-sea sand on a dig in Inverness. Similar deposits have been found in Shetland, Orkney, the Montrose Basin, the Firth of Forth and up to 50 miles inland. But the most profound effects must have been seen in Doggerland. It is likely that most of the low-lying subcontinent was drowned and reduced to a large island around the Dogger Hills and four or five smaller survivors to the south. Britain was at last isolated from Europe and, within a thousand years of the great tsunami, Doggerland had disappeared beneath the waves. But its people did not.

Those Scots who carry the M423 marker are almost certainly descended from the survivors of Doggerland or at least those ancestors who walked across the great plains on their journey westwards from northern Germany. Alongside the M284 and M26 markers from the Ice Age Refuges on either side of the Pyrenees, the children of the drowned subcontinent form one of the three founding lineages of Scotland. In all, these groups number around 150,000 of the modern male population – they were the earliest immigrants, those who first set eyes on Scotland after the ice.

The pattern of mitochondrial DNA is different. It appears that most European (and Scottish) mtDNA is even older than the Y chromosome markers, some of it reaching back well before the last Ice Age to around 40,000 BC. This was the time when the first bands of *Homo sapiens* entered Europe on the long journey out of Africa. The oldest types belong to the U5 and U8 groups of mtDNA and it looks as though they spread north and east once more from the southern Refuges around 12,000 BC. A large group, around 6 per cent of Scots or 300,000, belongs to U5. The most common mtDNA group is labelled H and it accounts for nearly half of the population. Group H appears to have arrived in Europe just after the peak of the last Ice Age and its variants again show a dispersal across Europe from the Refuges. The subgroups H1 and H3 in particular radiate from what is now the Basque Country across the continent, having begun to move rather later, around 9,000 BC. The ability to tell definitively which mtDNA lineages are related to which has only recently been possible, as the costs of reading all of the 16,569 letters of the mtDNA code has reduced. For that reason, this data is very general and it must be unlikely that the only origin of Scottish mtDNA lies in the south. Until more research on a much larger scale is done, the picture will remain blurred and piecemeal.

4

Harvest Home

✖

S URROUNDED BY THE grey and forbidding monuments of the
late twentieth century and hemmed in by its highways, the
remains of Scotland's most ancient building lay unnoticed
in an otherwise anonymous field. Beyond the main line and the
thunder of trains hurtling through the landscape from Edinburgh
Waverley to London King's Cross and beyond the insistent hum of
traffic on the A1 is the farm of East Barns, near Dunbar. On its
western boundary stands the Blue Circle cement works and to the
east is the Torness Nuclear Power Station.

When the cement factory applied to extend limestone quarrying
at East Barns, it was a condition that archaeologists were allowed
to survey and dig test pits in the land that was about to disappear.
And it was just as well they did for they came across something
spectacularly unexpected.

In the autumn of 2002, in a shallow hollow near the edge of
one of the fields, John Gooder and his team found the remains
of a prehistoric house. Scraping carefully with their trowels, they
slowly exposed the rim of a sunken floor. It was oval in shape and
measured 5 metres by 5.8 metres. Around its edges, circular pockets
of discoloured soil were recognised as large postholes. Diameters
varied between 25 centimetres and 55 centimetres and most of the
holes had been dug at an angle, strongly suggesting that the timbers

slotted into them had sloped inwards and formed a roughly conical shape.

It was clear from the outset that the excavators had come upon a very old site but no one could estimate how old until some organic remains were found. This was critical and, when some fragments of charred hazelnuts were identified, there was a chance that the age of the house might be estimated. When the results came back they were sensational.

The data told the excavators that the timber house at East Barns had been built and occupied around 8,000 BC, making it – and by some distance – the oldest building yet found in Scotland. The traces of shelters at Cramond (and at Daer, near Biggar, and Manor Bridge, near Peebles) were as old if not older but none of these could be described as houses, none could compare with the solid, even massive, structure which had been raised at East Barns.

Postholes of 55 centimetres in diameter had been dug to accommodate substantial timbers, probably the trimmed trunks of mature trees, and they will have achieved a roof apex visible from a distance on the flat ground by the seashore. Judging from the angle of the postholes, the shape of the house resembled that of a large tipi. The tree trunks sloped inwards so that that their ends met and interlocked, the weight of each making the heavy structure more secure. Perhaps they also lashed the apex with sinew cord to keep each member exactly in place. John Gooder and his team also found evidence for low walling around the sunken floor, probably turf, and their report speculates that the thick house timbers could have supported a heavy roof, perhaps also made from turf.

But it is the sheer scale and architectural complexity of the East Barns house which are so astonishing. Prehistorians have traditionally seen the pioneers who came north after the retreat of the ice as hunter-gatherer-fishers – small family bands who moved quietly through the landscape in pursuit of game and in search of the seasonal wild harvests of fruit, roots, nuts and berries. Living in shelters made from green stakes shoved into the ground or in caves or rocky overhangs, they left little but a gossamer trace of their passing, barely rustling the autumn leaves as they flitted through the virgin wildwood.

In 8,000 BC, life at East Barns cannot have been like that. In order to justify investing all that time, effort and skill in cutting the trees, digging the postholes, lifting the turf and making the big roundhouse weathertight, the band who built it must have insisted on some claim to own the land around it or at least to enjoy a set of customary rights to what it produced. And it has been conjectured that the very act of making the house, a clearly substantial and permanent dwelling, staked those territorial claims to any who might come to challenge them.

What was it about East Barns that persuaded the pioneers to make such a solid and impressive structure? To a prehistoric eye, such seaside places offered most possibilities – the timber house was near the seashore and all of its year-round bounty of fish, shell-fish and crustaceans, it lay at the edge of the dense wildwood and its wild harvest and it was near the Brock Burn, a reliable source of freshwater. Most important was a secure source of firewood. Hunter-gatherer-fisher communities needed to set and feed cooking fires all year round and inside the timber house they needed warmth and light in the long winter.

When supplies of firewood failed, it meant that pioneer bands were forced to move on, even leaving attractive sites like East Barns. To remain, for only a short period – a generation, perhaps – they will have needed to manage the woodland. The house construction shows that they could access and fell mature trees near at hand and it is likely that they undertook an organised and more widespread programme of woodcutting, leaving green boughs and trunks to season in caches before collecting them the following year. If distances gradually became too extended for a society without pack animals or wheeled vehicles, they will have been forced to transport logs in their curraghs. But, because all of the elements in woodland management were organic and perishable, no trace has been left of what must have been a vital activity.

Further inland, the shape of the landscape was attractive to pioneer hunting expeditions. The narrow, steep-sided valleys of Dryburn, Braidwood and Oldhamstocks in the Lammermuir Hills lay close at hand and they were perfect for what, in Scotland, is known as the 'drive and sett'. This style of hunting is very ancient and simple; it depended on a perimeter of beaters driving game before them through cover and forcing it towards the sett. There, hunters waited to shoot, spear or net the panicking animals as they fled before the beaters. The best sort of ground was known in Scotland as an 'eildrig' and this derives from an old Celtic root, *eileirg*, which meant 'a defile or a steep-sided narrow gorge'. Beaters tried to push game into narrow valleys like the Dryburn, Blackwood and Oldhamstocks so that those waiting at the sett could get close shots at their prey.

The drive and sett style was perfectly illustrated in the cave paintings of the Ice Age Refuges. The direct ancestors of the people at East Barns understood well how to use the landscape to bring down large animals. At Cueva de los Caballos in Spain, a stampede of deer and one aurochs, the huge wild cattle of prehistory, are being driven forward by men with sticks with flared ends (perhaps they cut branches with leaves to cause more panic) towards a sett of archers. Bows bent, they appear to have loosed arrows and hit at least two deer. It seems that these methods travelled north with the DNA of emigrants from the Refuges, in the founding lineage markers of M26 and M284.

Because this method of hunting needed numbers, it may be that the pioneers at East Barns used the techniques observed by archaeologists at the Ahrensburg reindeer slaughter. Perhaps hiding in the scrub and trees on the flanks of their valleys, they waited until their prey was abreast and close before firing arrows or throwing a spear. These strikes were unlikely to bring down a large deer or indeed an aurochs but, if they were badly wounded, it was only a matter of time before a hunting party caught up.

Clear traces of just such a blood pursuit have been found in the mud of the Severn estuary. At low tide, the blackened remains of an ancient forest are revealed every day and, between the stumps, archaeologists have found a remarkable survival. The fossilised footprints of a band of hunters in pursuit of a deer have been recognised. By careful measurement of the stride, it seems that they were stalking, setting down one foot slowly after another. Adult deer tracks have been found on the same path and, since such an animal could easily outpace men, it seems likely that the hunters were closing in on a wounded animal treading lightly through the Severn mud at the same time as the East Barns people were building their massive house.

In 8,000 BC, the shoreline of the North Sea lay about 300 metres away and that convenient proximity was a crucial factor. It meant that the East Barns people literally enjoyed the best of both worlds. A charred seal bone was found near the house and these creatures, so ungainly on land, made a very useful catch. More humbly, shellfish could be gathered on the wide expanse of rocky shelving exposed at low tide around Barns Ness. This was probably a chore for children and some of what they gathered will have been used as bait. In the spring and early summer, a more dangerous harvest was gathered in.

Further down the coast, where 150-metre cliffs rear up out of the North Sea, many thousands of seabirds nest on ledges. And off the coast of East Lothian, the Bass Rock and several smaller rocky islets are home to huge colonies. Their cliffs are white with guano. Both areas were reachable in a day in good weather by curragh and there were rich pickings for those with courage, skill and a steady head for heights. There is also some evidence to suggest that the East Barns people were adapted to manage the slippery ascent of the Bass Rock and elsewhere.

Eggs were gathered earliest in the year, no doubt placed carefully in backpacks by intrepid climbers. Later, they climbed to kill the young birds before they had fledged but, in both expeditions, hunter-gatherers will have had to fight off the shrieking assaults of the parent birds. Gannets, puffins in their cliff-top burrows and all manner of other seabirds were an important and salty part of a prehistoric larder.

The skills of Scotland's early peoples lingered for millennia. On the remote Atlantic island of St Kilda, men scaled the high cliffs for the young fulmar, either throttling them or clubbing them to death. And, like those of their ancestors, the bodies of the islanders had evolved. Always climbing barefoot, St Kildans' toes were markedly more widely spaced and longer to allow them to flex and grip the crevices and the rock face like hands. And their ankles were tremendously thick and powerful because they often had to bear an entire bodyweight on one leg or at an angle or both.

St Kilda was evacuated in 1930 but a hunt for young seabirds still takes place every year on another rocky Hebridean island. Sailing from the port of Ness at the northernmost tip of the Isle of Lewis, men hunt the guga, young gannets, on the uninhabited rock of Sula Sgeir. It is dangerous and unnecessary but the Lewismen go after the birds each spring in a remarkable cultural survival spanning ten millennia.

Continuities like the guga hunt may lead to a sense that our prehistoric ancestors were like us but without the technology and modern medicine. Certainly they looked much as we do, if a little shorter, a lot slimmer and less long-lived, but they did not think like us.

At its height around 16,000 BC, the pitiless white landscape of the last Ice Age had reached as far as the coastline of South Wales. But its spread was not uniform. In the east of England and the Midlands there was a polar desert and an enclave in north Yorkshire that appears to have remained ice free. The effect of this was simple. When the weather warmed, grass grew and the grazing herds moved north and it was to those areas they came first. Almost all of Britain's earliest traces of the hunters who followed the migrating animals, the men and women from the Refuges and from Doggerland, are found in clusters along the southern coasts of England. But there is one site, far to the north of these, that is perhaps most eloquent, that says something of how people thought and what they believed.

The northern English dialect word *carr* means 'a low-lying marshy place' and its best-known use is in the town of Redcar, 'the reed marsh'. Down the coast and to the south-west of Scarborough lies the Vale of Pickering and a clutch of 'carr-' names. The landscape is rolling, fertile farmland but it was once a huge *carr,* a marsh, and, before that, an even larger lake. After the Second World War, some prehistoric artefacts were discovered in a drainage ditch and, as a result, between 1949 and 1951, the site was systematically excavated. The finds turned out to be extremely informative.

Starr Carr was the first archaeological dig ever to be carbon dated and the results placed it very early in the human re-occupation of Britain. Around 8,700 BC, people were living at the edge of the vast lake and something of how they lived their lives has come to light.

Like the East Barns community, they had sophisticated carpentry skills and, using only flint tools, were able to trim and split timber (with wedges) to make a wooden causeway through the reed bed at the side of what the excavators called Lake Flixton. There was also a platform of more rough-andready birch boughs and brushwood to give some harder standing over the mud. As at East Barns, a great deal of flint debris was found, the spoil from making an estimated 14,000 tools. Flint knapping is close to an art and the depth of experience needed to examine a large piece and see precisely where to tap it so that the best blades and points can be made implies specialism. Perhaps there was a manufacturing centre at Starr Carr.

More striking was the discovery of 220 artefacts made from the antlers of red deer. It appears that the people at the lakeside were fond of venison and skilled at bringing down large animals. But the amount of antler bone found far exceeds what might be the proceeds of hunting by one band. It has been convincingly conjectured that Starr Carr was a place where prehistoric craftsmen made antler artefacts for other people. If that is so, then it implies trade or at least exchange of some sort and hints at a far more sophisticated economy than we might imagine.

Excavators also discovered that the reed beds by the lake had been regularly and deliberately burned. Why the people of Starr Carr did this is by no means clear. Perhaps they simply wanted to improve access to the water for their fishing coracles or perhaps they wanted to remove the dry reed wands to encourage green shoots to grow. These may have tempted browsing animals and made easier prey for waiting bowmen or spearmen. Waterside kill sites were common. It is also worth noting that while the peaty, anaerobic soil at Starr Carr has preserved a great deal, especially the antler artefacts, there is no sign of log boats, the alternative to curraghs. Hollowed-out log boats have occasionally been found at prehistoric sites but perhaps at the lakeside, with its wetland trees of birch and willow, there were few big trunks broad enough to make a boat. And, if the red deer were hunted regularly and in numbers, as the antler finds suggest, then the use of the hides of these large animals will have meant fewer seams to caulk in any skin boat.

Starr Carr and its great lake may also be seen in a wider setting, lying close to the western plains of Doggerland. In 8,700 BC, the subcontinent was visible from where Scarborough is now and may indeed have been easily reachable across a stretch of shallow water. Perhaps life at the edge of Lake Flixton resembled life on the shores of the Outer Silver Pit in Doggerland; perhaps those pioneers who lived in the east came to trade for antler artefacts. And perhaps they carried the same DNA marker of M423. It is still emphatically present in north Yorkshire and along the North Sea coasts of England and Scotland.

The most intriguing and evocative find at Starr Carr was a set of what were called red deer antler frontlets. Consisting of the forehead of a mature animal with the antlers attached, each frontlet had been fashioned in such a way as to fit on to a human head. Holes had been carefully bored through the skull so that thongs could be threaded and tied, probably under the chin.

These were headdresses, the ancestors of a long tradition whose most famous example was the Roman legionary standard-bearer, the signifier. These soldiers were reckoned to be amongst the bravest for they held and guarded the sacred emblem of the legion, its standard, and they often wore animal heads over their helmets and pelts down their backs. Before the reforms of Marius around 100 BC, signifiers wore bear, lion, wolf or horse heads. Much more recently, the Native Americans of the plains of North America were photographed wearing animal headdresses. Amongst the powerful Sioux nation, war leaders wore the familiar feathered war bonnet but others put on the skull of a bull buffalo, often with the animal's tail attached at the back. These headdresses were richly decorated and tied under the chin with thongs. Those with the buffalo horn war bonnet believed that they were imbued with the dignity and strength of the animal itself as they rode into battle.

In the half century since they were found, the red deer frontlets of Starr Carr have been interpreted and re-interpreted. Some historians imagined that they were used in deer hunts as a sort of disguise or decoy. This seems cumbersome and unlikely and the alternative view that the headdresses were made for ritual use is much more plausible. If the people of Starr Carr identified closely with the red deer, perhaps seeing themselves as the Deer Clan, they may have danced around the fire or performed a ritual hunt with men wearing the antlers. These are likely to have been heavy and difficult to manage without practice and, in a modern version of what may have happened on the shores of Lake Flixton, the men who take part in the deer-running at Abbots Bromley carry their set of antlers on their shoulders. Around this Staffordshire village, every September, six men each carry a set of huge, 12-point, reindeer antlers. They enact a ritual hunt. Even though the Abbots Bromley antlers are more than a thousand years old, there is no more than a very tenuous direct link but it may be that the rituals at Starr Carr had similarities.

Or perhaps the deer headdresses had an additional, much more brutal use. Because of the very real sense of small bands of men, women and children entering a virgin landscape at the end of the last Ice Age, we often take an Edenic view of the early millennia in our prehistory. As hunters stalked softly through the dappled shade of the wildwood and children gathered the bountiful natural harvest, it seems like an age of primal innocence. So few people migrated northwards from the Refuges or crossed from Doggerland to Britain that there was room and food for everyone, no cause for friction or conflict.

It cannot, of course, have been quite like that. In the beginning, Britain must have been sparsely populated and indeed hunter-gatherers needed extensive ranges to support even small groups. But pioneer bands must have encountered other groups or at least been aware of them – otherwise how did young people find marriage partners and how did the population grow? Contact must, on occasion, have produced conflict. Perhaps bands fought each other over territorial disputes. At East Barns (and a very similar house built at Howick on the Northumberland coast around 7,800 BC), there had been substantial investment in property, in making a sold and snug place to live and even a loose sense of ownership of the land around it would have prompted a need for protection and vigilance. No remains of large buildings have been found at Starr Carr but other forms of identification with territory may also have been powerful – the burial of the dead in what people saw as the kindred ground, the performance of rituals special to that place, the management and shaping of the landscape or disputes over women.

The ancestors and European contemporaries of the peoples who walked or sailed north to Britain after the retreat of the ice certainly possessed weapons and some of them made a record of their warfare. In three caves in eastern Spain, painters recorded fighting between bands. Two are perhaps better described as skirmishes but in El Molino de las Fuentes the artist has shown 15 archers on one side and 20 on the other. The dates of these cave paintings are by no means agreed. Some scholars believe that they are contemporaneous with the deer mask makers of Starr Carr, others that they are later. What is unarguable is what they show – war in prehistoric society.

Around 6,500 BC, there was a massacre in southern Germany. In a cave at Ofnet in Bavaria, archaeologists came upon two mass graves. They uncovered the grisly, even shocking evidence of extreme and indiscriminate violence – 38 decapitated skulls. Most were the heads of children under the age of 15 and some were younger than five. When the ferocious attack came, it appears that there were very few men amongst the victims and two-thirds of the adult skulls were identified as those of women. Perhaps many of their partners and the fathers of the slaughtered children were away on a hunting expedition. Perhaps the attackers knew that.

When the skulls were closely examined, those of the few remaining men had suffered the most grievous and repeated wounds, evidence strongly suggesting a futile attempt at defence. They may well have been old men – too old to go hunting, too old and outnumbered to protect their women and children. Several skulls exhibited cut marks on the crania and it seems that they had been scalped.

The Ofnet massacre is by no means unique. Elsewhere in Europe, skeletons retrieved from mass graves have given up the unmistakable signs of violent death at the hands of attackers.

Weapons capable of inflicting terrible injuries certainly existed in large numbers. Flint axes, arrows and knives can be razor sharp and one slash across an artery would have been lethal. Clubs seem to have been used at Ofnet and elsewhere and the likes of the archers on the cave walls in Spain could expect to wound and even kill with one shaft that found its target. Skeletons have been excavated with spear points embedded in their bones, usually the vertebrae.

If the men who tied on the Starr Carr headdresses led war bands, they will have acted as a focus for their warriors. Just as the Roman legionary signifiers were intended to be visible in the melee of fighting, the deer men could have acted as a rallying point. Or perhaps a prime value was aggressive display. Dressed in skins, perhaps wearing warpaint on their faces and chests, the men wearing the antlers might have presented a terrifying spectacle to any who threatened their people, the fearsome Deer Clan.

In the early prehistoric period, burials can sometimes offer a more lyrical sense of what people believed. It seems that when the hunting party returned to the scene of the Ofnet massacre, they buried their people with some ceremony and care. The skulls were covered with red ochre, a type of clay rich in iron oxide, and pierced red deer teeth and shells were also laid in the graves. This last sounds like jewellery and perhaps it belonged to some of the butchered victims. Red ochre was also used in a very early burial in South Wales. Dating to around 30,000 BC, in the time before the ice spread from the north, an adult male skeleton was found in Paviland Cave. His body had been covered in red ochre (mistaking his gender, early scholars called him the Red Lady of Paviland) and decorated with ivory ornaments. Several early prehistoric burials like this have been found.

It seems that the use of red ochre was symbolic, perhaps signifying blood and the fact that it had ceased to flow – or perhaps it simply conferred some sort of formality, perhaps it was the colour of high status. The use of the pigment has a long history and first comes on written record with the work of Jordanes, a sixth-century-AD Roman bureaucrat and historian. He wrote of the warlike Picts of Britain and their habit of painting themselves 'iron red'. And the tradition of warpaint was enduring in contemporary Ireland where bards sang of the deeds of the *Fer Dearg*, the 'Red Men'. Much later, the first Europeans to make landfall in Newfoundland encountered the Beothuk, a native people who wore red ochre on their bodies. It is the origin of the name 'Red Indians'.

The use of red ochre in prehistoric burials makes a direct link with the cave painters of the Refuges. They used a variety of ochres, mixing different sorts of powder to produce a wide variety of tints. Browns, yellows and purples as well as reds can be derived from iron oxide clays. The best quality red and yellow ochre in the world still comes from Rousillon in the south of France.

The use of symbolism in burials suggests sophisticated language, a clear ability to deal in abstracts. No belief system can be adduced from that, only a sense that our ancestors were capable of constructing one. Careful, even elaborate burial implies some consideration of continuity if not a life after death. Memory and a reverence for ancestors would certainly have been encouraged if a funeral had ritual and importance around it. If there was a prehistoric Heaven or Hell, then it is impossible to say much about what it was like. But there are burials that whisper of other worlds, places where the dead might live some sort of afterlife.

On the small Hebridean island of Oronsay, some time around 5,500 BC, skilled fishermen and hunters left a monumental record of their presence. On the coasts of this small T-shaped island, only two miles across at its widest extent, archaeologists have excavated five large middens. Now hidden under grass, these large piles of debris of shells, fish bones, deer bones, bit of tools and other assorted rubbish were found to be full of fascinating information.

In addition to many thousands of limpet and mussel shells, the bones of coalfish and cod were identified. These are not coastal species and they swim in deeper water, some way out to sea. Two sorts of deer bones have been recognised and since Oronsay was far too small and had little or no cover, these animals must have been hunted on larger islands such as Jura or on the mainland or both. What this means is good seamanship, a community that depended on its boats for food and for contact with other bands. Cod are big fish usually caught on lines and, to manage that successfully and safely, boatmen need a high standard of sea craft. The huge number of limpet shells found in the middens was likely to have been the residue of bait for these big fish rather than a rubbery part of the diet of the Oronsay people.

Venison offered variety, a break from fish, but the journey to Jura or the mainland was not made only for meat. The hides used for sea-going curraghs will have needed constant care, repair and replacement and, as at Starr Carr, the size of a deer pelt made it very attractive for that purpose. A set of bevel-ended tools fashioned from stone, bone and antler has been found in the Oronsay middens and these appear to have been used to make the hide softer and more pliable and more easily worked. This was certainly important for use on the hulls of curraghs but also for making clothes.

In the midden known as Cnoc Coig (it sounds better in Gaelic but all it means is Hill Five), evidence of a rare burial came to light. Human fingers had been carefully arranged on a seal flipper. This must be significant, a faint echo of how the Oronsay people thought about their dead. Were they bound for another life, to swim with the seals in the dark deeps of the world? The island lies on the edge of the Atlantic with its endless western horizon where the sun died each day and night fell. The vast ocean and its fathomless blackness may have seemed to the fishermen like an underworld, a watery version of the darkness of the caves of the Refuges, where their direct ancestors painted and worshipped.

More evidence of prehistoric violence came to light in the burials at Vedbaek on the Danish island of Zealand. A man's skeleton was found with a bone point lodged where his throat would have been. But at the same site an entirely different note was struck. When the remains of a child were excavated, archaeologists realised that a tongue-shaped piece of local stone had been deliberately placed in its mouth before the grave was closed. The idea of the burial of ancestors in the kindred ground would become widespread later but here was an example of apparent simplicity. Did the stone represent the land, the kindred ground, and its placing in the child's mouth signify how the bounty of the land fed its people?

A short distance away another grave hinted at something mysterious and very lyrical. A woman had been laid to rest with a necklace made from the pierced teeth of 43 different stags. Was she a great hunter or did someone who loved her bring down so many deer? In any event her burial speaks of high status, of a society beginning to stratify.

Beside the huntress lay an image of extraordinary poignancy. The body of a baby had been set on a white swan's wing and a small flint knife placed at its waist. And then both were dusted with red ochre. What could this have meant? It may well be that both the woman and her baby died in childbirth and it is not straining credulity to imagine a grieving father at the graveside. Did he believe that his baby's soul would somehow fly with the great birds? Swans could move between worlds; they were able to walk on the land, glide gracefully over the water and fly high in the sky. Like the Oronsay fishermen who laid the fingers on the seal flippers, the man may have been looking for the edges of another world, a shadow world where his baby and his woman might live on.

Perhaps the women came first, for once. Some time after 9,000 BC, the most profound transition in the world's history was beginning to happen. In Mesopotamia, the lands between the rivers Tigris and Euphrates and the uplands extending to the Mediterranean coast, the place once known as the Fertile Crescent and perhaps the inspiration for the myth of the Garden of Eden, men and women began to do more than manage the wild landscape. They began to farm it.

And it appears that mtDNA was important in the discovery and spread of the new ideas and techniques. As these started to ripple across Europe, two new markers arrived in Britain and Ireland. Dating can be problematic but it seems likely that, around or before 4,000 BC, women carrying mtDNA J1b came to Britain and Ireland from the south-west. Following a similar route to the one taken by the founding lineages who migrated from the Refuges, new people gradually made their way north up the Atlantic littoral. This marker gained an extra change or mutation to become mtDNA J1b1 and this is specific to Britain and Ireland. Approximately 3 per cent of Scots or 150,000 people retain it.

The second new marker also walked a well-worn path. Having crossed the Bosphorus from Anatolia, mtDNA J2a1 approached Britain along the great river valleys of central Europe, the Danube and the Rhine. It was a similar journey to the one taken by the founding lineage, M423, and it is this that makes the chronology imprecise. There is no doubt that the new mtDNA post-dated the earliest arrivals but it is impossible to pinpoint precisely when. Nevertheless, in the millennia after 6,000 BC, there is a powerful sense of seismic movement and of change taking place over the face of Europe.

In the Fertile Crescent, a revolution had happened and its radical new ways of thinking and working would change Europe and the world forever. Here were the origins of farming, the moment when people realised that they could sow and grow crops rather than simply gather a wild harvest – these transitions are, of course, recorded nowhere and their details entirely lost but a sequence of events and reactions may be recreated.

In the fertile region between the two rivers and the lands to the west, the climate was ideal for the rearing of cereal crops, the primitive wheat known as emmer, another type called einkorn and barley. A short rainy season was followed by a long, hot and reliable summer. All of the cereals began life as wild grasses and people began to select those with the most calorific seeds, store them and then sow them the next year. Those with small seeds or a bitter taste were discarded. Gradually those grasses with the highest density food parts, like wheat, barley and, elsewhere across the world, rice and corn were bred selectively and yields increased. Longer stalks were also encouraged as a source of forage for animals and straw bedding for human beings.

There were eight founder crops in all, the staples upon which farming was built. In addition to emmer and einkorn wheat and barley, early farmers cultivated lentils, peas and chickpeas. And, for its oil and fibre (ultimately used to make fabric), they also grew flax. The care of these crops and fields was almost certainly women's work. With the very recent mechanisation of farming, the ancient role of women as fieldworkers has been forgotten but the traditions of the bondagers and plough-followers of rural Scotland reach far back in time.

Women had another motivation to rear cereal crops. In hunter-gatherer societies children took a long time to wean. As their teeth developed slowly there was no suitable substitute for breast milk and, as a consequence, it could be four years before a child could subsist on the adult diet of roots, nuts, fruit and meat. The effect of this was a low birth rate. While mothers are lactating it is unlikely that they will conceive, and birth intervals of four years or so meant a birth rate of only three or four children at most for the short lifetime of each fertile woman. As at the swan's wing burial at Vedbaek, it seems that mothers regularly died from complications in childbirth and in later prehistoric societies, samples of populations show few living beyond their mid-twenties.

When cereal crop production reached a critical mass in the Fertile Crescent, populations are likely to have increased very rapidly. The ground seeds of wheat or barley could be mixed into a warm porridge and it was perfect fare for weaning small children off their mothers' breasts. That, in turn, reduced the length of the birth interval, perhaps halving it to two years or even less. By itself, this shift in cultural practice had a dramatic effect on population and eventually prompted movement away from the overcrowded fields of the river valleys. Beyond them, new lands waited to be cultivated.

At the same time as women tended crops and children, men probably took a leading role in looking after animals. Domestication was the vital and complementary second element in the farming revolution. Species that were sociable, docile and meaty were chosen. Cattle, goats, sheep, pigs and chickens were all reared and selectively bred. Virtually omnivorous and capable of producing large litters, pigs were the only animals bred solely for meat. All of the others had secondary uses, contributing milk, wool and eggs. And incidentally, only the pig, or rather the wild boar, was indigenous to Britain in the early prehistoric period. Sheep, domesticated cattle, chickens and goats all crossed the sea to take their place as a familiar part of our farming landscape.

All dogs descend from wolves and they appear to have been domesticated early. Perhaps orphaned cubs attached themselves to hunter-gatherer bands and found a role as guard dogs. Early herdsmen seem to have trained pastoral dogs not so much for the sophisticated gathering skills shown so attractively by Border collies and other breeds but as a means of detecting by scent and warning by voice of the approach of a wolf pack. More recent hunter-gatherer societies also valued dogs as companions, and when Australian Aboriginals describe a cold night sleeping out in the bush they talk of a 'three-dog night'.

Horses were first subdued and ridden across the steppes of central Asia and their speed transformed communication. Around 4,000 BC, riders began to travel long distances and one historian has described horses as the prime vehicles for the westward transmission of the family of Indo-European languages. Also used as pack animals able to carry far greater weights than human beings, they transformed trade both in bulk and reach. It took many centuries for horses to replace or supplement oxen as draught animals (yoked to a sled or a cart, ponies are prone to imagine they are being chased by what they are supposed to be pulling) but, when it did happen, the economic and military effects could be spectacular. In 1,674 BC, a chariot-driving people known as the Hyksos swept down into Egypt, where horses were little known, and established themselves as pharaohs.

Successful cereal growing brought many more changes in its wake. Most immediate was the invention of secure storage. Once grains were harvested and dried (or charred to prevent germination in a warm climate), most of the crop needed to be kept safe so that it could be used as required. Granaries were dug or erected and lined to keep out damp, rodents and other pests. But the most important innovation was the fired ceramic pot. These developed quickly, could be of varying sizes and weights and were portable. Providing a good stopper sealed the top properly, no rats or mice could nibble their way in.

There were, however, damaging drawbacks to this new way

of life. As men, women and children lived alongside animals, diseases jumped the species barrier. Humans caught tuberculosis from cattle or influenza from chickens or pigs. Tending crops and keeping animals off them was also an unrelenting business and farming communities tended to work harder and longer than their ancestors. They also appear to have been generally less healthy and shorter. The average height for men declined from 5 ft 10 in. to 5 ft 6 in. and for women from 5 ft 3 in. to 5 ft 1 in. Nevertheless the increased birth rate drove up numbers and people were forced to move on.

Latitude was a prime consideration. When populations in the Fertile Crescent grew beyond the capacity of the land to support them, small groups, perhaps bands of younger children, were compelled to seek new fields and pastures. But, if they wished to live by planting crops and rearing animals, they could not travel far to the south or to the north. Cereal crops, and wheat in particular, will only flourish in certain climatic conditions. Wheat will not grow in the tropics or where the weather is too cold or wet. The same is broadly true for domesticated animals – most do better in a temperate climate where there is good grass for them to graze. And so, when groups of emigrants began to move out of the Fertile Crescent, they walked eastwards towards India and China or westwards into Europe and the lands around the Mediterranean Sea.

By 5,500 BC, farming had advanced up the Danube valley, reached the Hungarian Plain and acquired an archaeological profile. Having cleared areas of woodland and scrub, communities built a particular style of timber longhouse. These could be modest at 10 metres in length or spectacular at 40 metres. To store and consume what they grew, the new farmers made a new pottery. Known in German as *Linearbandkeramik* or LBK, it was decorated by lines incised in the clay before it was fired. While these groups did cultivate crops (and few will have abandoned hunting and gathering entirely), it seems that the LBK users were primarily cattle herders. From Hungary, the culture of cultivation and stock rearing spread quickly to the rest of continental Europe, advancing at a rate of a kilometre a year.

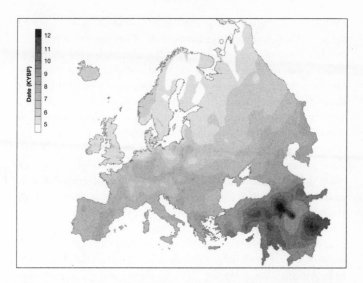

*The contour map shows the dates of early Neolithic farming sites in Europe.
The earliest sites are in Anatolia and Mesopotamia (in darkest shading).
The two routes, around the Mediterranean and up the Danube basin can be
appreciated from the slightly darker shading in these areas
(i.e. earlier arrival of farming).*

In a rare dry and hot summer in 1976, archaeologists took
advantage of the good weather and took to the air to do a pho-
tographic survey of parts of Aberdeenshire. On the banks of the
River Dee, at Balbridie, crop marks suggested the outline of a very
large building. Measuring 26 metres in length and 13 metres broad,
it looked as though it was one of a type constructed in the Dark
Ages. Excavators began work on what they imagined was the sort
of hall described in the famous Anglo-Saxon poem *Beowulf*.

When organic matter was recovered, carbon dating produced a
startling result. Far from being the feasting hall of Anglian warriors,
Balbridie had been built some time around 3,900 BC. Its appearance
must have made a tremendous impression, for no structure on this
scale had been seen in Britain. In fact, it was wider than any longhouse built in Europe. But Balbridie was not alone. Similar crop

marks have been detected at another site on the opposite bank of the River Dee. And, at Claish Farm on the River Teith in Perthshire, another great timber hall was raised a century later. Excavators found traces of internal divisions, rooms where people may have slept or worked or stored household items. At one end of the hall at Balbridie, a concentration of pottery finds suggested that someone had been making it. But there were no traces or indeed likelihood that animals had been brought in to overwinter in a byre area, even though there seems to have been plenty of room. The internal divisions and supports for the huge roof both at Balbridie and Claish could not have allowed animals to enter or be accommodated. It seems that the hall builders were making a home, a place where an extended family or a lineage might all live and in relative comfort.

At Balbridie, a staggering three quarters of a million grains were found in the postholes – mostly emmer wheat, bread wheat and barley. Clearly cereals were cultivated on a large scale for so much to have been dropped or lost. The hall builders were farmers but, before they felled trees and began to split timbers to make the walls of their house, they will have taken much care over the choice of a site. Another similar structure has been detected near Kelso in the Scottish Borders and it illustrates the priorities of these remarkable people best. Like Balbridie and Claish, the hall at Whitmuirhaugh Farm stands on a river terrace and it is surrounded by undulating, fertile and, above all, well-drained land. And there is a large and dependable water source near at hand. The rich, dark soil by the Tweed has been seen as very valuable throughout history. In the seventh century AD, an Anglian lord built another hall nearby and, in the twelfth century, the monks at the new abbey at Kelso kept Whitmuirhaugh in their own 'dominium' as a grange to be worked by lay brothers because it was so productive.

These four halls can be seen as the earliest farmsteads yet found in Britain – another has been discovered in Kent but it appears to be slightly later and others have come to light in Derbyshire and Oxfordshire. They are on a monumental scale and seem to appear suddenly in the landscape with no obvious local antecedents predating them. But across the North Sea there existed several similar examples. Timber longhouses were built on fertile river terraces in the Rhine and Seine valleys and in northern Germany and associated with them were finds of LBK pottery, cereal production and evidence of stock rearing, cattle in particular.

The conclusion seems patent – the great halls in Scotland and elsewhere in Britain were raised by immigrants, people who sailed the North Sea or more likely the English Channel (by 3,900 BC Doggerland had probably entirely disappeared) with their animals, their seeds, their tools and their skills. It may well be that they crossed by the shorter sea routes, then made their way up the North Sea coast and inland up the arterial rivers Tweed, Forth and Dee. It is important, however, to bear in mind the accidents of survival and discovery. Between Kent and Scotland the remains of many more prehistoric timber halls may have been lost or ignored.

It used to be thought that farming migrated across Europe as an idea rather than something brought by waves of people on the move. But it turns out that both interpretations may be correct. People did move but perhaps not far and not in large groups. Farming depends not only on having the seeds of founder crops and domesticated animals in tow but also on knowing what to do with them. As its techniques spread westwards across Europe, small bands and their dependants may have travelled and made new settlements in good places. Then their descendants moved again and passed on their skills. Only two of the eight founder crops, flax and barley, grow wild outside of the Fertile Crescent and the others will have had to be transported by migrating farmers.

Balbridie, Claish, Kelso and the English halls were built on what had become an island by 3,900 BC and the rapid progress of farming across the continent at one kilometre a year had to accelerate and make a tremendous leap. And it appears to have done so. Farming arrived in Aberdeenshire and elsewhere virtually fully formed and that does not speak of gradual transmission and the adoption of new ideas but of the arrival of new people. Who were these pioneer farmers, where did they come from and do their descendants still live in Scotland?

DNA evidence is not yet conclusive, only very suggestive. Studies of mtDNA have tracked markers such as mtDNA 2a1 and J1b1 arriving in the fifth and fourth millennia BC by two routes – the Atlantic littoral and along Europe's great river valleys. And there are indications that a third and fourth, T1a and K2a, came to Scotland at around the same time. Together these four lineages account for about 10 per cent of the modern population. Research carried out on female skeletons found at the sites of LBK communities of farmers in Germany and Hungary has identified another marker, N1a, which is now rare, present in only 0.2 per cent of all Europeans. It is found in Scotland and there is an extremely close match with a lady from East Lothian farming stock. She may indeed be a direct descendant of one of the LBK skeletons. But these are small numbers. If farming did change Scotland profoundly, does that mean there are only faint traces of the people who created that revolution left in Scotland's DNA?

Partly because the traditional migration routes to Britain along the Atlantic coast and through the central European valleys did not change, dating the arrival of new DNA markers can be difficult. Unlike mtDNA, for which dates can be inferred reasonably well, dating the origin or expansion of new Y chromosome DNA markers or the lineages they define can be difficult. This is because the markers used to estimate the dates mutate or change at very different speeds and the actual rate of change is poorly understood. They can be seen as a palimpsest of patterns, one laid on top of the other.

However, a new Y chromosome marker does look as though it came with the pioneer farmers. The M269 marker defines a group known as R1b and it is predominant in Scotland with 70 per cent or 1.75 million men carrying it and more than 110 million in Europe. It does originate in south-west Asia but, beginning with the tiny groups who repopulated Europe 40,000 years ago, people have been walking to the west from that direction for millennia. What persuades geneticists and historians that R1b was the farmers' marker is its very rapid multiplication and spread. How did that happen? How did M269 get to be so common so quickly?

The overwhelming likelihood is that it came to Balbridie, Claish, Kelso and elsewhere and, as farming was rapidly and successfully adopted, immigrant pioneers took native women as partners (perhaps polygyny was the norm) and their Y chromosome marker began to expand exponentially. With the cereal harvest and the milking of domesticated animals supplying alternatives to breast milk and the length of the birth interval decreasing, farming communities quickly became large. In some parts of Eastern Europe, villages began to grow to extraordinary sizes. In the foothills of the Carpathian Mountains and in the valley of the River Bug, some were colossal extending to a thousand acres and with 10,000 inhabitants. Nothing like that happened in Scotland but farming was a hard-working way of life, needing many hands, and, as in rural Scotland until the nineteenth century, having many children was a practical blessing. And it seems that the pioneers needed big houses.

The arrival of M269 represents the most influential immigration in all our history. The people carrying it may not only have changed our way of living profoundly, they also had the most significant effect on our gene pool. There are telling differences between genders, between the proportions of mtDNA and Y chromosome DNA, and these may point to an ancient pattern of cultural change. It appears that female-line ancestry is predominantly pre-farming while male ancestry dates mostly from the coming of the farmers. At such a distance and with no other historical evidence to underpin such an assertion – aside from comparisons with other eras and other patterns of colonisation – it looks as though incomers took native women as their partners and on some scale.

Many sub-lineages from the M269 group have been recognised, some of them arriving in Scotland later (sometimes from Ireland, where they had existed since pre-farming times). Other Y chromosome groups show a descent from the first farmers. M172, M201 and M35, for example, developed out of the earliest peoples to come to Scotland and change the landscape and the way in which it was used. These three lineages are most common in the south of the country, clearly showing the direction of approach. However, in deeper time, their origins can be traced in different locations. M201 is most common in the Caucasus Mountains while M172 seems to have arisen in the Fertile Crescent. Most fascinating is the journey of M35. They are the most divergent Y chromosomes to be found in Scotland and, in fact, took a different route out of Africa from all other Scottish lineages. They did not descend from the band of pioneers who crossed The Gate of Tears to modern Yemen. The ancestors of those Scotsmen who carry M35 probably left Africa thousands of years later and followed the course of the River Nile as it flowed northwards. They then crossed the Sinai Peninsula and reached the Levant, modern Syria and Lebanon, where they dispersed amongst the communities who were beginning to domesticate animals and cultivate crops. This extraordinary lineage is much more distantly related to all other Scottish lineages and it shares ancestry with many African types deep in its remarkable past.

New evidence published in 2015 both complicates and confirms this patchwork picture of the arrival of farming. A paper published in the journal *Nature* by Wolfgang Haak, of the Austrian Centre for Ancient DNA, and others promises to revolutionise current thinking about the peopling of Europe after the end of the last Ice Age. Having analysed ancient DNA taken from 69 individuals who lived between c.6,000 BC and c.1,000 BC, Haak and his colleagues now believe that there were two major migrations of farmers into Europe. The first took place between c.6,000 BC and c.5,000 BC, and the new people came from the Near East, modern Syria, Lebanon, Israel and Jordan.

The new research shows that this migration was followed by a period when the native hunter-gatherer populations were resurgent. Then there followed a second, massive migration about 2,000 BC. This came from a different direction, from the steppes that stretch between the northern shores of the Black Sea to the Caspian. Known as the Yamna or the Pit Grave culture, these people were pastoralists who grazed and tended their herds and flocks on the wide grasslands. Characteristic of these societies were kurgans, grave mounds that were often large and could be seen from afar on the vast horizon of the steppe. Many of the Yamna herders moved westwards around 2,000 BC and Wolfgang Haak and his fellow authors have produced evidence from ancient DNA to show that the steppe migrants were immediately and dramatically successful, replacing about 75 per cent of the ancestry of central Europeans. Their impact in Scotland remains unclear – but it is highly unlikely that there was no impact.

What makes this discovery even more important is its effect on thinking about the theories around language shift. Some believe that the progenitor of almost all European languages, what is known as Proto-Indo-European, came from the east, but now it seems that it arrived in the mouths of the Yamna. New farming and herding techniques needed new terms to describe them and perhaps they were first used on the steppe grassland, what is now eastern Ukraine and southern Russia. Their displacement of 75 per cent of central European ancestry will have ensured that their language became quickly dominant. And perhaps it is no coincidence that the earliest archaeological definition of a Celtic culture derived from two type sites in Central Europe, at La Tène in Switzerland and Hallstatt in Austria.

If indeed 70 per cent of all Scottish men are descended from immigrants whose DNA originated in the Near East less than 10,000 years ago, then perhaps Scotland's most famous farmer was unconsciously thinking of his ancestry when he ended his great anthem 'A Man's a Man for a' That' with the lines 'That Man to Man, the world o'er,/Shall brothers be for a' that'. It seems that we are cousins if not brothers of men who live many thousands of miles to the east. In the twenty-first century the Islay farmer felt a sudden and astonishing tug of history when he had his DNA analysed. His variant of M172 was so specific that, in the male line, he was more closely related to his Iraqi and Syrian genetic cousins than he was to his fellow islanders.

The significance of the arrival of farming in Scotland is simple. Before Balbridie, Claish and Kelso, our ancestors depended on the landscape and its flora and fauna for their survival. To be sure, they did manage it and no doubt tried to conserve and husband where they could but it was the land that fed them. Like the cave painters at Lascaux they lived in an environment dominated by the natural world. After the building of the great halls, men and women tried to use the land to feed themselves as they modified crops and bred preferred characteristics into the animals they reared. Floods, droughts, disease and natural disasters apart, the peoples who lived in Scotland began to behave like masters of their world.

5

Sunrise and Moonset

✖

WHEN THE EXODUS bands began their long walk northwards out of Africa around 70,000 BC, it is likely that
those men, women and children spoke to each other in
the same language. When they crossed strange horizons and came
to settle in strange lands, they used familiar words to describe what
they saw. And just as their DNA and their two female and two male
lineages are the ancestors of all the world's non-African DNA, did
their language give rise to all the rest of the world's languages? This
difficult question is not a matter of mere intellectual curiosity or an
opportunity to erect entertaining hypotheses. Languages describe
the people who speak them, their worlds, their shared beliefs, their
customs and habits. But before written records, the only course
open to historians willing to deal with this centrally important question is to work backwards in an attempt to reconstruct these ancient
ways of understanding life.

In 1903, a Danish scholar, Holger Pedersen, proposed a very
early language he called Nostratic. The name derived from the
Latin *nostrates*, meaning 'fellow countrymen', and it described what
Pedersen believed to have been the common speech of people at
the end of the last Ice Age. Nostratic was argued as the ancestor
and basis of all the languages of Europe, Asia and North Africa.
Much comparative research was done in the 1960s by two Soviet

scholars, Vladislav Illich-Svitych and Aharon Dolgopolsky, and they demonstrated many surprising similarities. A Nostratic dictionary was compiled. Known as the Moscovite School, the Soviet scholars were suspected of having an ideological motivation lurking in the shadows behind all their theorising. Totalitarian states have an infamous tendency to manipulate history to suit the status quo and, as a federation of tremendous linguistic diversity, from Leningrad to Vladivostok, the Soviet Union needed all the unity it could muster. Despite these real or imagined taints, the notion of Nostratic has outlived the breakup of the USSR and is once again taken seriously.

The Urheimat or original homeland of this primal language is now thought to have been in the Fertile Crescent, and the engine that powered its spread, both westwards into Europe and eastwards into Asia, was the development of farming. As ideas and techniques fanned outwards, so the new words needed to describe these innovations moved with them. It was not only a matter of vocabulary – the names of plants, animals, tools and chores, new ways of saying things about new methods, new ways of living were required.

As a few farmers and the knowledge of farming penetrated to Europe, their revolution overturned the pre-existing social organisations of hunter-gatherer-fishers. This did not happen everywhere and precisely how society was ordered can only be a matter of speculation. History is rarely tidy. But even if new people were in a small minority, what they had to say about food production and the language they said it in must quickly have become dominant. If the hall builders of Balbridie, Claish and Kelso carried the farmers' marker of M269 and it rapidly reproduced itself in the new fields by the banks of the Dee, Tay and Tweed, then the new speech community will also have expanded very quickly.

That is not to say that language extinction took place. Across Europe, Asia and North Africa the broad family of Indo-European languages is certainly diverse and different geography and climate may well have encouraged the survival of the different vocabulary needed to describe it. To drag in a cliché, Scots have more words to describe rain than Greeks do.

Few people had more words to describe everything than Sir William Jones. An impoverished but brilliant scholar at Oxford University from 1765 to 1768, whose family roots were in Anglesey or Ynys Mon, Jones was a hyperpolyglot. By his teens, he was fluent in Greek, Latin, Hebrew, Persian and Arabic and he understood the basics of Chinese. By the time he died at the young age of 47, this extraordinary man had complete command of 13 languages and was conversant in a further 28. Paying his fees from work as a translator and tutor, Jones took a law degree and became a circuit judge in Wales. In 1774, he enjoyed a moment on the world's stage when, as a young lawyer, he found himself in Paris negotiating with Benjamin Franklin in an attempt to find a peaceful solution to the problems simmering in the American colonies. Nine years later, William Jones was appointed as a judge on the Supreme Court of Bengal and, with his linguistic gifts supplying immediate access, he became immersed in and entranced by Indian culture.

It was language itself that fascinated the young lawyer and he gave a seminal lecture to the Asiatic Society in Calcutta. Here is the most famous passage:

> The *Sanscrit* language, whatever be its antiquity, is of a wonderful structure; more perfect than the *Greek*, more copious than the *Latin*, and more exquisitely refined than either, yet bearing to both of them a stronger affinity, both in the roots of verbs and the forms of grammar, than could possibly have been produced by accident; so strong indeed, that no philosopher could examine them all three, without believing them to have sprung from some common source, which, perhaps, no longer exists; there is a similar reason, though not quite so forcible, for supposing that both the *Gothic* [Germanic] and the *Celtic*, though blended with a very different idiom, had the same origin with the *Sanscrit*; and the old *Persian* might be added to the same family.

Sir William Jones was by no means the first to notice similarities between Sanskrit and European languages. An English Jesuit missionary of the late sixteenth century, Thomas Stephens, thought that Greek and Latin had clear links. Sanskrit is the liturgical language of both Hinduism and Buddhism and, while modern versions are spoken, its status is as a language of antiquity – like Latin and Greek. Another westerner in sixteenth-century India saw more modern parallels when he found that if he counted in Italian, Sanskrit speakers understood. Filippo Sassetti noted how *sette* was like *septa*, *otto* like *asta* and *nove* like *nava*. But what William Jones could see was a much broader and more profound series of connections. As a hyperpolyglot, he heard the common chatter of the Indo-European languages in his head and, perhaps because of that, he found it easier to learn more and more of them.

Basic words were shared. Father in English is *pater* in Latin and Greek, *fadar* in Germanic languages, *pitar* in Sanskrit and *athair* in Gaelic. In farming, there is an extensive lexicon of similarity – for grain, *granum*, *gran*; for milk, *mulgeo* ('I milk'), *melg*; and, for honey (except in English), the root *mel-* or *mil-* is shared by dozens of languages across Europe and Asia. In English we hear it in mellifluous, meaning 'sweet sounding'.

While disagreement, differing interpretation and constant revision are amongst the staples of academic discourse, one principle in all this is clear. Whether or not it is known as Nostratic or Proto-Indo-European, it appears that Europe's and Asia's languages are more than linked. They did indeed have a common ancestor. It was spoken on the long walk out of Africa, the crossing of the Red Sea into Arabia and when the pioneer bands at last arrived in the Fertile Crescent.

While some scholars have argued powerfully for the spread of Indo-European from what is now the Ukraine, carried by the horse-riding warriors of the Kurgan culture, it is much more likely to have been broadcast by the spread of farming. While there is no doubt that military elites can impose their language and have done so throughout history, there are better reasons to seek more peaceful means of transmission. Farming moved simultaneously east and west, it needed language for it to be successfully adopted and, over a vast area of the world, peoples were motivated by self-interest rather than compelled to learn it.

Before the coming of written records, language shifts and developments are very difficult to date. They tend to be processes rather than events. Sanskrit is thought to be the oldest recorded, used as early as 1,500 BC in recensions of the religious texts known as the *Rig Veda*, a collection of hymns. Farming reached India much earlier and evidence for the cultivation of cereals in the sixth millennium BC has been found in the Indus Valley.

At the same time, the hall builders had begun to plant the dark loam of the Hungarian Plain. If – and of course this is a consider-able leap – both waves of cultural change, farming and the language that described it, washed both east and west at the same rate (and the distances between the Fertile Crescent and both locations are almost exactly equivalent) of one kilometre a year, then it is likely that dialects of a related language began separate development in the sixth millennium BC. Shared DNA, the M269 and M172 markers, testify to this.

That aside, much harder evidence exists for a fundamental social change in the north of Britain in the fourth millennium BC. Large, impressive and essentially mysterious monuments began to appear in the landscape. Many were monuments to the dead and they were built by communities of farmers to commemorate their ancestors and perhaps to consolidate their hold on their land. Often these sacred places had humble beginnings and often they are much older than first anticipated.

When the new town of Glenrothes was planned for central Fife after the Second World War, it was envisaged that its economy would centre around a large new colliery, the Rothes Pit. But this closed in 1961 because of flooding and the town was forced to import new industry such as electronics and to rely on the expansion of established manufactures like paper making. By the 1970s, Glenrothes was expanding and more housing was planned. A prime site was earmarked at Balfarg but the existence of two immense standing stones (one at 6 ft 7 in. and the other at 5 ft 3 in.) prompted the immediate involvement of archaeologists. These impressive stones once stood at the entrance to a henge. But more surprising was the discovery of much earlier traces of religious activity. Later excavations in the 1980s found a series of pits around the henge at Balfarg. Carbonised barley grains were noticed inside a sherd of pottery and they were carbon dated to 3,600 BC. The pits contained burnt material, pottery and bones and these fragments seem to have had sacred significance. They were the slight beginnings of a series of religious monuments built over millennia on the site.

Around 2,900 BC, a wooden henge was raised using 16 timbers, probably trimmed tree trunks. Because standing stones survive in situ, especially large ones, we assume that henges were always constructed by heaving them into place. But in fact most of these circular enclosures were wooden and, at Dunragit near Stranraer, there stood a mighty circle of timber, much larger than the stone-built Ring of Brodgar or the Stones of Stenness on Orkney.

Late on in the third millennium, Balfarg's timbers were replaced by two concentric stone circles and, around 1,900 BC, a grave was dug in the centre. The corpse of a young man, someone of high status, was placed in it and a flint knife and a beaker set beside him.

From the postholes of the earlier wooden circle at Balfarg, pottery was lifted out that had once contained a hallucinogenic drug. Black henbane is poisonous but, in the right quantity, it can also produce hallucinations. Other easily available plants such as fungi could have a similar effect. It seems that the rituals of the henges involved trances, almost certainly dancing, incantations, music, perhaps song and processions. If the circles of stones and trees marked off a sacred precinct often hidden by a bank and screens, a sanctum sanctorum, then it will have been entered with some ceremony and drama.

The needs of ritual could be spectacular – and dangerous. When the evening sun falls behind the jagged ridge of the Langdale Pikes, it backlights them black against the blue sky. While tourists bustle and tills ring merrily down in the valleys, the mountains of the Lake District still rise above in majesty. In the fourth millennium, they were also seen as magical. High up on the summit ridges, along vertiginous ledges, there are ancient quarries of volcanic stone. Even though workable deposits of the hard, fine-grained rock exist lower down and in much more easily accessible places, the prehistoric miners preferred to climb close to the peaks of Pike o'Stickle and Scafell Pike. There, as the wind whistled and the rain blew in off the Irish Sea, they hacked out stone to make axes.

This was an immense undertaking. The screes still visible on the lower slopes of Pike o'Stickle are largely the waste that clattered down from the high quarries. Before being carried downhill, axe heads were roughed out to minimise the weight and the screes at Langdale are reckoned to be the debris of more than 45,000. Once down on the lower ground, the axe heads were polished with abrasives to a dark lustre. Then they were taken all over prehistoric Britain. The readily recognisable geology of the volcanic rock of the Langdale Pikes has allowed archaeologists to trace a distribution from Aberdeenshire to Kent with concentrations in Cumbria, East Yorkshire and Galloway.

What made the Langdale axe heads so sought after and so magical? Perhaps they were a symbol (many seem never to have been used) of the clearance of the wildwood and the making of the fields needed for arable farming. It seems likely. But the deliberate choice of quarries at such high altitudes and in such dangerous places implies another magic – the involvement of sky gods in the power and beauty of the axe heads. Weather mattered a great deal to farmers and maybe they had begun to direct their prayers and anxieties skywards.

All over Britain and Ireland, monuments were rising. In the late fourth millennium and early third millennium BC, the dead were commemorated in a series of very striking sacred and ceremonial sites, the layout of many of which seems to be linked to the transit of the heavens. Stonehenge, Avebury and the stone circles on Orkney are only the most famous but their beauty and mystery must not be allowed by the accident of survival to blind us to the undoubted existence of monuments in all of the regions between. These took many forms and were often complex and always enigmatic. Any attempt to recreate the detail of the rites that took place in and around them cannot succeed. It is like trying to reconstruct the nature of Christian belief by looking at the ruins of Britain's great abbeys.

What the farmers' monuments do clearly signal is a very different sort of society. The carbonised grains found at Balfarg were the product of great labour. Clearings had to be made, roots ripped out and the ground turned so that it could act as a seedbed. Sometimes this was done by burning back the wildwood and archaeologists have found evidence of seeds being planted directly in the ash of extinct fires. Once shoots began to show, animals had to be kept off until harvest time. Then the cycle had to be repeated, often with newly made clearings since cereals exhaust the soil quickly. One perceptive historian has reversed a central question. Instead of asking how hunter-gatherers progressed towards the higher levels of an agricultural society, he wondered what persuaded them to abandon a relatively secure, well-provided and stress-free way of life.

Farming was hard work but, in good years, it was also rewarding. Populations rose and more hands could do more. It was that and the surplus production of food and the ability to store and preserve it that enabled the great monuments to be built. Farming allowed communities to work on projects which did not produce food. More precisely, the cultivation of cereals, and wheat in particular, freed people to think about other things, about abstracts, an afterlife, the gods.

Across the world, farmers came to cultivate three different staples – rice, maize or corn and wheat. Europeans preferred to grow wheat where possible and to make bread with its flour and this choice or accident had profound historical consequences. With an average of 70 per cent carbohydrate and 12 per cent protein, wheat is very nutritious. And it was partly because its protein content is much higher than that of rice or maize that Europeans grew taller and more muscular than East Asians or South American natives.

There was another advantage – one that can be seen at Balfarg, the Langdale axe quarries and in the dazzling variety of the prehistoric monuments of Britain and Ireland. Unlike rice whose paddy fields need constant maintenance throughout the year, wheat requires only two periods of intensive commitment. The clearings need to be turned and the seeds sown in the spring and, in the autumn, the harvest needs to be reaped and brought safely in. There are of course other chores needing attention but cereal farmers have always had time when their crops are growing ripe, time to build spectacular monuments, time to become political, to fight and to think about worlds beyond.

Counting became a key skill for farmers. In the wildwood and on the seashores and estuaries, hunter-gatherers cannot have been ignorant of basic arithmetic. The catch after a drive and sett hunt needed to be counted and divided and, for fishermen, the ritual of tipping out the creels to see exactly how many fish should be shared lasted into the nineteenth century in Orkney. And guga hunters were used to dealing with the large numbers of young birds they piled into their boats. But stock farmers had to be able to add and subtract often and this common cultural requirement is reflected in both language and the evolution of numbers.

While green stalks of wheat slowly ripened in the clearings and the fields by the riverbanks, they were less likely to be trampled or eaten by herds of cows, sheep or goats. From spring until autumn, most stockmen (and they were likely to have been men and boys) drove their beasts upcountry on the ancient journey of transhumance. In order to rest the pasture of the valley bottoms that had been churned to mud and grazed bare in the winter, flocks and herds were led up the hill trails to the high ground. There summer pasture was fresh, juicy and plentiful, good for cows and ewes making their first milk. In Scotland, place-names remember the old journey. The Lammermuir Hills were at one time the Lambs' Moors, the high ground where ewes grazed and suckled new lambs.

In spring and autumn squalls, shelter could be hard to find and the remains of buchts and shielings can still be seen in many places over Scotland's hills. Some are very old. And often they were no more than a rickle of stones set against the prevailing wind – a place where ewes and their lambs could find a warm bield. The small cattle of the first farmers were shaggy beasts and, with their calves, they made their way down into the deans and declivities to wait out the bad weather. Much milk was made by the nursing animals and there are long and attractive traditions of young women joining their shepherds, 'laughin' and daffin' at the yow milkin''.

Few farmers will have had large herds in the fourth and third millennia BC but counting became especially important at calving and lambing time – and when their beasts roamed free over unfenced grazing. Others will have had certain customary rights to hill grass, what became recognised as the common land in the medieval period. Communities will have asserted these rights and, if there was contact with neighbours, then it was important to be clear about how many beasts there were and who owned them.

One of the earliest counting systems on the historical record – and one that is familiar, if indecipherable, to many from the formal inscriptions of dates on neoclassical buildings – is Roman numerals. Like our Arabic numbers, they look abstract but, in fact, they are no more than hand signals, deriving from fingers, thumbs and the shape of the human hand – one (I) is a straight line or one finger held up and two (II) and three (III) are simple progressions, five (V) is the nick between index finger and thumb and four is one finger held up to the left of the V and six one finger to the right of it and so on, while ten (X) is a cross of both index fingers representing all the fingers and thumbs. Decimal counting (such a contentious innovation in Britain from 1970 onwards) only came about because we have ten fingers (or eight and two thumbs). Fifty (L) is the nick between thumb and index finger again but this time with the index finger pointing upwards and C for a hundred is the same arrangement but with the finger and thumb crooked. It is simple and was probably the way shepherds reckoned how many sheep and lambs they had before and after lambing and perhaps how they indicated the number to neighbours.

The early words for numbers are fascinating and clear indicators of how close a relationship exists between the Indo-European group of languages. And these words are very old.

Such is the power of three, for example, that it is virtually the same word from Britain to India and all points between – three, *tri, drei, trios, tres, tri, tri* and *tri*. Celtic, Germanic, Romance and most Asian languages count in a very readily recognisable way – up to ten. Numbers mattered enormously and there needed to be agreement or at least mutual understanding. It is perhaps no accident that there are such similarities over such a wide area.

But, as languages settled and localised, traditions began to vary.

Sanskrit operates in units of tens and the teens are simply additions – one and ten for eleven and so on. Roman numbers became conglomerates of units, fives and tens. This system was a simple adaptation of how people used the fingers and thumbs of both hands. I was one finger held up, two was II, three III, five was the angle between index finger and thumb, while four was that with I on the left, the minus side. Ten was crossed index fingers. English became entangled in dozens (particularly when currency was formalised) but perhaps the most interesting and the oldest means of counting in Britain and Ireland is still to be found in the indigenous Celtic languages. Welsh, Gaelic and Irish have all preserved what is known as a mixture of Base 20, Base 10 and Base 5. For example, fifteen in Gaelic is *coig deug* or five and ten, while sixty is *tri fichead* or three twenties and eighty is *ceithir fichead* or four twenties. Base 20 counting involved toes as well as fingers and it survives here and there in newer languages. The French use of *quatre-vingt* for eighty harks back to the centuries when Gaulish, a cousin of Gaelic, was spoken.

The Old English word 'tithe' meant a tenth and it is a relic of a common form of taxation. Into the early modern era, the clergy and the nobility were entitled to a tenth of the produce of farmers in a parish or an estate. There is no direct evidence of tithing or any other sort of taxation in the fourth or third millennium BC but the very fact of building so many complex and impressive monuments insists both on an organised society and one that had begun to stratify into hierarchies. The henges at Balfarg required a directing mind both to envisage them and to make them happen. Aristocracy, perhaps even kingship, certainly leadership of some kind, is embodied in an increasing emphasis on elaborate burial, the veneration of the dead and of ancestors. The man laid in his grave at the centre of the Balfarg henge was not a nobody.

As society developed into a hierarchy, those at the top naturally felt a need to protect their elevated position and they required taxation in kind to achieve that. If a king, a queen or a lord began to exert control over a community, then enforcement of some kind cannot have been far behind. It seems very likely that food rents were ingathered to support a group of men whose duties were military. Perhaps it is too much to call them warriors or a war band.

More than a thousand years before Balfarg was built, the cities of Sumer in Mesopotamia had developed aristocracies, high priests and kings. Their agriculture had grown very sophisticated with year-round cropping, irrigation, the employment of a specialised labour force and a huge annual cereal crop. Sumer means the 'Lands of the Lords of Brightness'.

Around 3,000 BC, farming in Scotland was not like that – it was much less intensive, less extravagantly productive and it built no city states. But similar social dynamics are very likely to have been at work. It is stretching credulity to imagine that society was not organised even if on a much less sophisticated level than the bright cities of Sumer. There were prehistoric lords, perhaps even warlords, in Scotland and it may be that their monuments are the henges, the barrows and the dazzling variety of prehistoric constructions still visible in the landscape.

Few are as impressive as the stone circle at Easter Aquhorthies in Aberdeenshire. The Gaelic part of this place-name recalls its ancient purpose – it means 'Prayer Field'. Built in the fourth millennium BC, there are more than a hundred circles like it in north-eastern Scotland. Small, between only 18 and 25 metres across, they have a recumbent stone sitting between two large uprights. Set on its side the recumbent stone is like a table or an altar and it is always to be found on the south-western perimeter of the circle. Beyond it there is, in every case, a distant horizon, never closer than a kilometre away. At Easter Aquhorthies, the singular peak of Bennachie can be seen clearly, not occluded by any other feature in the landscape.

While assiduously avoiding any hint of New Age quackery, archaeologists have come to the view that the time of the midsummer full moon was significant in the design of this sort of stone circle. Flanked by the two tallest uprights, the recumbent stone framed the low moonrise against the clear skies of the summer solstice. Without modern light pollution, the rays of the full moon at that time of year can make night seem like day. Perhaps it was the moonshine rather than the moon itself that was important in whatever ceremonies took place inside the circle of the old Prayer Field.

The recumbent stone circles are a local phenomenon, a striking reminder that Scotland's prehistoric culture was not uniform but very varied. Clustered in Aberdeenshire, Kincardine and Buchan, these sorts of monuments were often found close to each other. Were they the focus of an estate or a group of farms belonging to a kindred, a proto-clan ruled by a priestly chief who officiated at ceremonies inside the circle built by his people or their ancestors? Perhaps they were the descendants of the hall builders who came to the banks of the Dee.

Archaeologists have come across another intriguing find particular to the north-east and associated with the recumbent stone circles. Intricately and very skilfully carved stone balls, most no larger than an orange, seem to have been gifts or tokens of some kind. Certainly they appear to have no practical value or purpose. Perhaps they have something to do with authority, like the orb and sceptre in coronation ceremonies. Or maybe they acted as tokens – symbols of amity and cooperation between communities, something given when marriage alliances were agreed or new arrangements over grazing or access to water or some other agricultural matter were made. The Langdale axes (only five of these have been found in the land of the recumbent stone circles) may have been used in the same way. Many of the balls have been carved in facets with grooves between them and in such a way as to allow them to be tied with cords and possibly hung where all could see them. It should be remembered that just as the austere ruins of classical Greece and Rome were usually brightly painted and adorned with flowers and offerings, so the henges of the prehistoric peoples were likely to have been decorated and made festive.

Recent and remarkable discoveries in Orkney reinforce this notion. At what has been called a Neolithic cathedral, a large and impressive building discovered between the Ring of Brodgar and the Stones of Stenness, archaeologists came across two stone slabs with unmistakable remains of paint on them. They talked of 5,000-year-old Dulux. Carefully trowelling away the earth, they recognised colours – red, yellow and orange. Ochre powder had probably been mixed with a binding agent to make it stick to the stone surfaces. Egg whites have been used in this way for at least two millennia and many of the great frescoes of the Italian Renaissance were painted on plaster in egg tempera.

The archaeologists who discovered the painted stones also brought to light a group of remarkable buildings, the most impressive and complex monuments built by the early farmers yet to be discovered in Europe. Two massive walls, more than 110 yards long, had been built right across the narrow isthmus from between the shores of Loch Harray and Loch Stenness. And at 14 feet in height, it was a formidable barrier, screening completely a complex of no fewer than ten temples, the largest extending to an area of 80 feet by 65 feet. Like the hall at Balbridie, it was immensely impressive from the outside but had a restricted, darkened interior with room for only a small group of people. Whatever the nature of the rituals celebrated on the Ness of Brodgar, the scale of the temples strongly suggests that they were the business of an elite and not designed to be seen by a large group.

The great temple compound at Ness of Brodgar is the largest non-funerary structure in Europe and seems to have been the focus of a hallowed landscape, a huge religious complex between the lochs at Harray and Stenness, a place where people processed, worshipped and made music in rituals now completely lost. And these landscapes and the individual monuments in them were the first of their kind in Britain. The creation of sacred enclosures that marked off the inside from the temporal outside, the holy of holies from the world beyond them and the circles that became known as henges, were invented in Orkney. Brodgar and Stenness were raised before the stones were erected at the more famous circles at Avebury and at Stonehenge. It appears that a messianic figure or group of figures may have created a new way of worship on Orkney, a directing mind(s) who invented rites that involved whole communities in massive communal work programmes – even though they appear to have been celebrated by an elite. At Brodgar, stones were definitely brought there from different parts of the archipelago as though each community was contributing and there is a sense of the megaliths as mute, monumental representatives. And there is also the sense of a highly organised, confident, productive society governed by a dynamic, innovating elite – at least for the period when the great monuments rose in the landscape.

At other archaeological digs, small pots have been found to contain pigment and perhaps they were used in some way to add to or touch up paintwork. While it is dangerous to generalise from isolated examples, it is worth recalling the skills of the cave painters of the Ice Age Refuges and the possibility that they passed on an almost entirely lost art along with their DNA.

Other theories about the stone balls of Aberdeenshire have included an unlikely use as weapons. Perhaps they were symbolic maces in a similar way in which the Langdale axes represented practical axes.

If summer moonshine was what fascinated the farming kindreds of the north-east, it was winter sunshine that caught the imagination of prehistoric architects in Orkney. Built around 2,700 BC, Maeshowe is a large chambered tomb inside a mound. The drystane corbelling is marvellously made and the structure largely intact. Approached through a low passageway (probably constructed like that to compel an attitude of reverence), the main chamber contained the bones of the dead, the ancestors. It appears that corpses were usually defleshed before interment in tombs like Maeshowe. At Balfarg, what is thought to have been a mortuary enclosure has been recognised. The recently dead were apparently laid on high platforms so that scavenging birds such as ravens could peck their bones clean. A fence around the platforms was intended to prevent predators from getting at the bodies and dragging parts away. The preservation of the whole skeleton seemed important – even though it was not always kept together in a chambered tomb.

The low entrance faced a range of hills across a flat stretch of the Orkney landscape and around the time of the winter solstice the rays of the setting sun shone directly through it and into the main chamber of the tomb. To ensure the correct alignment, the architects of Maeshowe must have experimented with a prototype structure of some sort. In any event, the effect is both moving and spectacular. The sun penetrating the tomb in exactly the way anticipated by its builders somehow draws us closer to them. Here is the great Orcadian writer, George Mackay Brown:

> The most exciting thing in Orkney, perhaps in Scotland, is going to happen this afternoon at sunset. In few other places, even in Orkney, can you see the wide hemisphere of sky in all its plenitude. The winter sun just hangs over the ridge of the Coolags. Its setting will seal the shortest day of the year, the winter solstice. At this season the sun is a pale wick between two gulfs of darkness. Surely there could be no darker place in the be-wintered world than the interior of Maeshowe.
>
> One of the light rays is caught in this stone web of death. Through the long corridor it has found its way; it splashed the far wall of the chamber. The illumination lasts a few minutes, then it is quenched. Winter after winter I never cease to wonder at the way primitive man arranged, in hewn stone, such powerful symbolism.

A similar effect can be seen at the huge burial complex of Newgrange in the north of Ireland. There the entrance passage is aligned to the rays of the midwinter sunrise at the solstice and they shine through an opening above the lintel known as the light box to illuminate stone panels in the burial chamber on the shortest day of the year. Scholars have calculated that the measurements involved in constructing these monuments were not absolutely precise. At Maeshowe, the sun's rays penetrate the tomb for more than 40 days each winter.

It is often claimed that the great prehistoric monuments were architectural calendars or sun-or moon-dials but this must be unlikely. In a general way, the seasons did of course matter to an entirely rural population but the times to turn the earth in the fields, to plant and reap or when it was best to drive stock up the hill trails – these varied each year. The weather, the temperature and the temper and condition of the animals themselves all had a bearing on the turning points of agricultural year. They still do. Farmers do not need elaborate monuments to advise them – they look up at the sky, sniff the air and pick up a handful of soil in their hands. The impressive monuments – by their very scale – meant a great deal to all who lived around them but they cannot have been practical in that way. Moonshine, sunshine and the mysteries of the heavens and things we can only guess at lay at their hearts. They were built for celebration not calculation.

Some slight sense of this can be read in fragments of a remarkable document. Around the year 325 BC, a Greek explorer and geographer visited the British Isles. Travelling north from the mercantile colony of Massalia (modern Marseilles), Pytheas made his way up the Atlantic coasts and landed on the Isle of Lewis, probably coming ashore where Stornoway Harbour is now. There he took measurements, calculating from the height of the sun that he had reached 58° of latitude north. Pytheas set down the details of his journey in a periplus, a working guidebook known as *On the Ocean*. While the original manuscript has been lost, extracts have survived in the work of later classical geographers. Following Pytheas, Diodorus Siculus wrote of an island in the far north visited by Greeks and where Apollo was worshipped 'in a notable temple which is adorned by many votive offerings and is spherical in shape'. This is almost certainly a reference to an ancient moon cult.

So far, so vague. But then Diodorus quotes what sounds like a clear echo of a passage from *On the Ocean*:

> They say also that the moon, as viewed from this island, appears to be but a little distance above the earth . . . The account is also given that the god visits the island every nineteen years, the period in which the return of the stars to the same place in the heavens is accomplished . . . At the time of this appearance of the god he both plays on the cithora [lyre] and dances continuously the night through from the vernal equinox until the rising of the Pleiades . . .

The 19-year cycle of the moon is a vital clue to the precise location Pytheas was writing about. It is only in regions close to 58 degrees latitude north that the moon 'dances' along the horizon without setting, and it does this every 18.6 years between the spring equinox and the old Celtic festival of Beltane on 1 May. Scholars have conjectured, with a great deal of confidence, that this passage is a reference to Callanish, the majestic alignments of standing stones on the Atlantic coast of Lewis. They were hauled upright around 2,800 BC and, in a rough cruciform, they point to the four quarters of the compass. Pytheas certainly took a measurement of latitude on Lewis some time around 325 BC. Did he cross the island to visit the ancient stones, did he listen to stories about the ceremonies enacted by the Old Peoples who raised them? It seems likely.

If the moon and its rays were central to the purpose of Callanish, then that clearly chimes with the intentions of the builders of the recumbent stone circles of Aberdeenshire. By the time Pytheas could have seen Callanish, the stones would have seemed less monumental. Peat had formed around them but its encroachment had not diminished their memory amongst those people the Greek traveller met.

Towards the end of the third millennium BC, the genetic make-up of Scotland was beginning to divide. A British variant on the European marker M269, known as S145, could be said to be the most emphatic signal of the Celtic language speakers of the British Isles. It is found in England, Wales, Scotland and Ireland and it is almost certainly characteristic of those farming communities who may have spoken early forms of Celtic languages in the centuries around 2,000 BC. A recent large sample of Welsh men showed not only how stark the different distributions of R1b S145 could be but also how they have persisted for millennia. More than 48 per cent of all Welshmen are in the classic Celtic Y chromosome haplogroup. This compares to only 15 per cent of all men in neighbouring eastern England. Offa's Dyke turns out to have been a genetic as well as political frontier. A similar division is detectable in Scotland, although the differences in distribution are not so stark. The R1b S145 haplogroup is present in Glasgow and the south-west at a frequency of 31 per cent of all men, but this declines to 22 per cent in Edinburgh and the south-east. However, after 2,000 BC, new markers and striking changes can be detected.

Burials became different. Bodies were being interred as individuals rather than as one of many in the chambered tombs. Inside stone-lined graves known as cists a dead person was often laid on their side in a foetal crouch. Millennia before, Neanderthal burials had been like this. Soil analyses of the floors of cists sometimes show that the graves had been strewn with flowers, the beginning of a long tradition of wreathes at funerals. One bloom was especially valued for its fragrance. Meadowsweet is a tall-stemmed white flower found in abundance in damp places. Still very common in Scotland, it has both a powerful and sweet scent and also useful medicinal properties. Meadowsweet's roots contain salicylic acid, an ingredient of aspirin and when small quantities are chewed it can relieve pain and headaches.

These properties of the plant have been understood for millennia but it was for its use as a floor covering that meadowsweet became known. Mixed with bracken and straw its fragrance kept household smells manageable and it was famous as a favourite of Elizabeth I of England. But the name betrays an older use. Meadowsweet derives from 'mede-sweet' or an agent for making the honey-based alcoholic drink mead even sweeter. Deposits have been detected in prehistoric drinking vessels found in cist graves in Scotland. It seems that decorating burials with flowers and drinking a valedictory toast to the deceased are both ancient cultural habits.

What was new some time around 2,000 BC were the vessels containing mead or ale. A new style of pottery began to appear. Known as beakers or bell beakers after their usual shape, these were reddish-coloured and decorated by pressing twisted cords into the soft clay before it was fired. They seem to have been an import from Europe where earlier examples have been found over a wide area. It used to be thought that their appearance in Britain marked the arrival of new people, the Beaker Folk. Perhaps it did – to an extent.

Pots and rounded food bowls were not the only items placed in graves. In a cist uncovered at Culduthel in Inverness, very beautifully made archery equipment was found. In addition to a large beaker pot and eight delicate flint arrowheads, the excavators came across something new and extremely significant. When archers pull back a taut bowstring to loose off an arrow, the string can whip against the wrist of the hand holding the bow. In the Culduthel grave lay a bone wrist guard dating to around 2,000 BC – and it had been studded with rivets made of gold. This find is amongst the earliest evidence for the arrival of metalworking in Scotland. But in the south something even more spectacular came to light at a very famous site.

In the summer of 2002, a new school was planned for the small town of Amesbury, near Stonehenge. Before building work began, archaeologists were called in to check that nothing of importance would be obliterated. They found two burials dating to 2,300 BC and the associated grave goods were startling. The body of an older man had five beaker pots arranged around him, and no more than two had ever been found in a single burial before. Like the Culduthel body he had a stone wrist guard and 15 arrowheads but there was more. Three knives made from copper had been laid in and, most significantly, a smooth 'cushion' stone, a small tool last used by early smiths for working metal.

In the second burial lay the body of a younger man who was certainly a relative. Both skeletons had foot bones that are normally separate fused together in the same way. The younger man had two very similar gold objects that at first were thought to be earrings. They were, in fact, probably hair decorations since they were shaped like a spiral or the hoops of a barrel and would have been plaited into pigtails. An analysis of the older man's teeth showed that he had travelled a long way to his death at Stonehenge. He came from central Europe, probably somewhere in the northern foothills of the Alps. His son or nephew was however local and his teeth showed that he had drunk the water of the chalky Wessex landscape when he was growing up.

It is very dangerous to generalise from one example, no matter how impressive. Nevertheless the discovery of the Amesbury Archer is hard evidence for the physical transmission of new technology from Europe to Britain and Ireland. In the beginning of the era of metalworking very few will have had the necessary skills and they would have been unlikely to make more than a tiny impression with their DNA, given the considerable numbers of people living in the British Isles at that time. And the two burials show how the skills were passed on, perhaps in secret, to family members. As a smith, a man with the magical talents of turning base ores into gleaming metal, the Amesbury Archer was accorded a sumptuous funeral. The grave goods suggest a great occasion with much pomp and ceremony, the passing of a person with tremendous prestige. And his origins are equally suggestive. At the beginning of the first millennium BC, the metalworking Hallstatt culture of central Europe was flourishing. It was seen by some as the fount of later Celtic Europe, the place where famous artefacts were made. Was the Amesbury Archer a product of an earlier concentration of these new skills in the same place?

The evolving differences in British and Irish DNA make a straightforward and perhaps timely point. New variants of M269 such as S21 and S28 were being brought into the east of Britain. In Scotland, in particular, geography was of course a factor. The Drumalban Mountains, called by St Adomnán the Montes Dorsi Britanniae, the Mountain Spine of Britain, were difficult to penetrate except through a few passes. The main artery, the Great Glen, ran north-east to south-west and a sea journey around the north risked the open Atlantic. Three clear variants in Y chromosome DNA are detectable in the modern population. In eastern Scotland, S21 is reckoned to be the marker of 30 per cent of men in Stonehaven and 27 per cent in Morayshire. In the Hebrides, only 7 per cent of men have that marker and, in Oban, it is 14 per cent. A similar bias can be observed in the south. In the Lothians and the Borders, there are many more with S21 than in Galloway and the same is true for the S28 and M253 markers, although the differences are somewhat less pronounced. With S145, a western marker, the pattern is reversed.

In England, a similar east/west divide has been seen in Y chromosome sampling despite the fact of a lesser geographical divide. Why has this split come about? The answer may well lie in the workings of seaborne trading networks, the sorts of contact that brought the Amesbury Archer to Stonehenge. Because we are a land-based culture used to roads and cars, we can forget the pull of historical gravity towards the sea. To repeat, it was not a barrier in prehistory but a highway.

Artefacts as well as people moved between Europe and Britain as trading contacts intensified. Not only had the culture of the burial of important individuals with their grave goods spread widely, from the Mediterranean to the North Sea and from the

Hungarian Plain to the shores of Biscay, but the characterisation of these members of an elite as archers also became more common. No doubt bows, now long decayed, were laid in graves as well as quivers of flint-tipped arrows and wrist guards. And no doubt these had evolved as general symbols of status. Perhaps the gear of the archers had come to be universally understood as the attributes of a great warrior or a great huntsman. It must, in some cases, have been purely symbolic. Experts examining the Amesbury Archer found that he had suffered a debilitating knee injury that would at the very least have caused him to limp and probably given him a great deal of pain. He was very unlikely to have hunted or fought in battle but his metalworking skills accorded him high status.

DNA sampling reinforces an intertwined sense of two distinct seaborne trading networks in Britain and Ireland. In the west, the emphatic presence of S145 appears to mirror mercantile contact. Distinctive pots known as maritime bell beakers were first made in the region around the River Tagus in Portugal and the tradition of bows and arrows in graves may also have originated there. By 2,500 BC, this cultural package had spread north to the Morbihan area of southern Brittany and the mouth of the Loire. This area became a centre of production and exchange not only for bell beakers but also other valuable items such as axes, flints, daggers and lance heads. From Morbihan/Loire the beakers filtered down the French river valleys to the Mediterranean coast and eastwards to northern Italy. To the north, contacts were made with Wessex, Ireland and Atlantic Scotland.

Now, it appears that S145 also travelled these trading routes. The marker probably originated in southern France and northern Iberia and people carrying it came to Ireland and western Scotland. This was not a wave of migration but a series of small movements over time, probably in the millennium between 2,500 BC and 1,500 BC.

On the eastern coasts of Britain traders were just as active. And with their goods came new people and their DNA. The marker S21, so common in eastern Britain, is mirrored by a high incidence of the same one in northern Germany and the Low Countries, particularly in Frisia, the necklace of islands on the north-western coast of Holland. Also more common in eastern

Britain, S28 originated in the lands around the Alps, south-east France and northern Italy and then spread across what is now Germany. Because of the relative imprecision of the Y chromosome molecular clock, geneticists have occasionally urged caution in comparing these samples of modern populations. Could S21, M253 and others not have arrived much later in Britain, especially in England, with the coming of the Anglo-Saxons and Danes after the fourth century AD? But in Scotland at least the notion of a more ancient east/west divide is much more robust because it is observed clearly in areas where there was little or no settlement by Anglo-Saxons. In Moray and Aberdeenshire, the incidence of S21 is very high in the male population and that of S145 rather low. No doubt the Anglo-Saxons brought S21 and other markers across the North Sea once more, strengthening the gradient of genetic types across England, but they were present in England long before.

Memories of migrations before written records are rare and, where they occur, the details are often muddled and mythic but sometimes suggestive. The Tower of Hercules is a remarkable – and unique – building. A lighthouse first raised by the Romans in the second century AD, it stands 60 metres high on a commanding cliff-top location in Galicia in northern Spain. It looks out over the Atlantic and its light was visible from ships far out in the ocean. Around the base of the tower winds the remains of a ramp built for oxen to pull the cartloads of wood needed to keep the light burning brightly.

The location of the Tower of Hercules is puzzling. Geography prevents it from providing a guide into safe harbour for boats approaching from the southern or eastern coastlines of the Iberian Peninsula. Instead it can be best seen from the northwest, from the open ocean. Did ships not hug the coasts for once, not daring to sail out of the sight of land? Instead did they swing out wide into the Atlantic to avoid the notorious squalls of the Bay of Biscay? Was there therefore more direct contact between Galicia in northern Spain and Brittany, even Britain and Ireland? It seems so. The construction and maintenance of the lighthouse was so laborious as to discourage the idea of a monument built for anything other than a practical purpose.

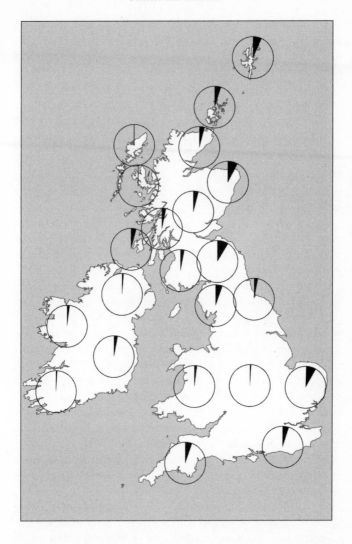

The frequencies of the S28 Y chromosome group are shown across the British Isles using pie charts. Up to 3,000 samples were used to create this map.

Until the twentieth century, the Tower of Hercules was known as *Farum Brigantium*, the 'Light of Brigantia', an old name for the nearby port of Betanzos. The earliest mention of it is in the work of the Roman historian, Paulus Orosius:

> At the second angle of the circuit [circumnavigating the Iberian Peninsula around Cape Finisterre], where the Galician city of Brigantia is sited, a very tall lighthouse is erected among a few commemorative works, for looking towards Britannia.

Galicia is the most north-westerly region of Spain and, just as Wales, north-western Scotland and Ireland still do, it has pungent Celtic traditions. Its name recalls them. Galicia has the same derivation as Gaul, Galatia and the French name for Wales, Pays des Galles, and there are persistent links with Ireland in particular.

In his chaotic, occasionally fantastical and fascinating *History of the Britons*, the ninth-century priest, Nennius, wrote of waves of colonisation from Galicia to Ireland. And, in a collection of poems and prose from a long and ancient oral tradition first written down and edited in the eleventh century, these traditions are much amplified. *An Lebor Gabala Erenn*, literally *The Book of the Taking of Ireland*, more usually *The Book of Invasions*, tells the tale of King Breogán, the first ruler of Celtic Galicia. It was he who commanded the Tower of Hercules to be built. And it had to be tall enough for his sons to see a distant green shoreline. Breogán encouraged his son, Ith, to take ship and sail north to Ireland so that they could take it for themselves. An attractive fable, much later it persuaded Galicians to erect a huge statue to Breogán close to the ancient lighthouse.

Behind the myth-history of *The Book of Invasions* lies a much less dramatic narrative of contact and migration. But the detail was very beguiling. Edmund Spenser, the author of *The Faerie Queene* and an Elizabethan civil servant in sixteenth-century Ireland, had read that the country had indeed been colonised by people from northern Spain but not by Celts and he offers an attractive derivation of a familiar name:

The chiefest nation that settled in Ireland suppose to be Scythians . . . which first inhabiting and stretching themselves forth into the land as their numbers increased named it all of themselves Scuttenland which more briefly is called Scuttland or Scotland.

The Scythians were real enough although they did not, in historical reality, penetrate anything like so far west. Their gorgeous, bloody and golden culture flourished in what is now the Ukraine and the Black Sea coasts between the seventh and the fifth centuries BC. Horse-riding Scythian war bands certainly made their way into the heart of Europe and their ponies grazed the lush Hungarian Plain. Their martial prowess was famous, their war gear thought lethal and their cavalry skills much feared. And Edmund Spenser's derivation of the name of Scuttland or Scotland is at least intriguing.

The Elizabethan antiquary and historian William Camden was impressed when he wrote of the Irish tradition of descent from a Scythian prince, Fenius Farsaid. He had settled briefly in Iberia before crossing the sea to take Ireland:

[T]o derive descent from a Scythian stock cannot be thought in any way dishonourable seeing that the Scythians, as they are most ancient, so they have been the conquerors of most nations, themselves always invincible, and never subject to the Empire of others . . .

What people believed is too frequently judged against modern standards of objective truth for often it is as important to an understanding of history as what they actually did. William Camden was only repeating a long tradition when he wrote of the Scythians as never having been subject to others. And it appears that he did not doubt it. This belief may have been what prompted Abbot Bernard of Arbroath to include these warriors in what must be Scotland's most famous, most quoted historical document. What most remember from the Declaration of Arbroath is the stirring passage about Scotland's independence from England: 'For so long as there shall but one hundred of us remain alive, we will never give consent to subject ourselves to the dominion of the English.'

But often overlooked is the opening of the letter to Pope John XXII. Abbot Bernard's purpose in composing it was of course to bolster Robert de Brus's claim to the Scottish throne and to underpin its legitimacy by summarising Scotland's long history and its origins as a separate nation:

> Most Holy Father and Lord, we know and from the chronicles and books of the ancients we find that among other famous nations our own, the Scots, has been graced with widespread renown. They journeyed from Greater Scythia by way of the Tyrrhenian Sea and the Pillars of Hercules, and dwelt for a long course of time in Spain among the most savage tribes, but nowhere could they be subdued by any race, however barbarous. Thence they came, twelve hundred years after the people of Israel crossed the Red Sea, to their home in the west where they still live today. The Britons they first drove out, the Picts they utterly destroyed . . . and, as the historians of old time bear witness, they have held it free of all bondage ever since. In their kingdom there have reigned one hundred and thirteen kings of their own royal stock, the line unbroken by a single foreigner.

It made little difference that Robert de Brus was certainly a foreigner – his family were originally French, probably from the Cotentin Peninsula. What mattered was the sense of continuity and Abbot Bernard's unblushing co-option of *An Lebor Gabala Erenn* was probably seen as a strength rather than a plagiarism.

The Irish had resisted English domination as energetically as their cousins, the Scots. Bernard's short list of indigenous enemies, the Britons and the Picts, more than suggests that he was writing of the gradual takeover of the whole of Scotland by the Gaelic-speaking Argyll kings of the west, the macAlpin dynasty. But that is to anticipate events.

The journey from Iberia is firmly supported by DNA evidence but the idea of the Scots coming out of Greater Scythia, the lands to the north and east of the Black Sea, through the Mediterranean and the Straits of Gibraltar – surely that is completely fanciful?

The Crowd rejoices as Scotland wins a Grand Slam at Murrayfield in 1990

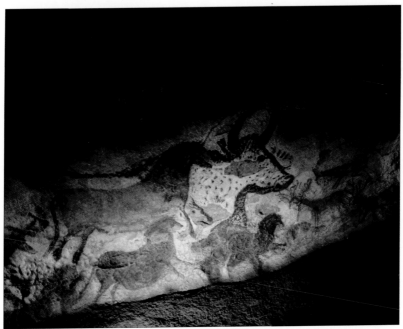

Wild cattle, horses and deer thunder across the walls of the cave at Lascaux
(Getty Images)

Right. What Ötzi might have looked like

Below. A hunter–gatherer–fisher harpoon from the island of Oronsay (© Dumfries and Galloway Council. Licensor www.scran.ac.uk)

Sir William Jones, the great linguist
(© National Portrait Gallery,
London)

A pottery vessel from the henge at Balfarg
(© National Museums Scotland. Licensor
www.scran.ac.uk)

The recumbent stone circle at East Aquhorthies in Aberdeenshire
(© Crown Copyright reproduced courtesy of Historic Scotland)

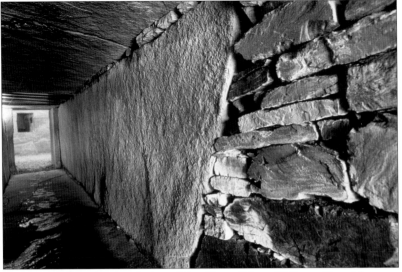

The interior of the chambered cairn at Maes Howe, Orkney (© Crown Copyright reproduced courtesy of Historic Scotland)

The mound of Maes Howe, built some time around 3,000 BC (© Crown Copyright reproduced courtesy of Historic Scotland)

The centre of the standing stones at Callanish, Isle of Lewis (Liz Hanson)

Grave goods found at Culduthel, Inverness (© National Museums Scotland. Licensor www.scran.ac.uk)

A mummified body found at Cladh Hallan, South Uist (© Mike Parker Pearson)

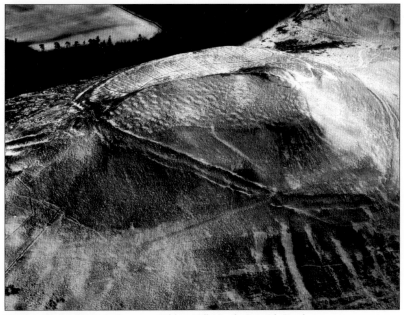

The hundreds of hut platforms on Eildon Hill North, Scottish Borders (© Royal Commission on the Ancient and Historical Monuments of Scotland. Licensor www.scran.ac.uk)

Above left. The Aberlemno Stone, showing the Battle of Dunnichen, 685 (© Crown Copyright reproduced courtesy of Historic Scotland)

Above right. The magnificent Ruthwell Cross (© Crown Copyright reproduced courtesy of Historic Scotland)

Middle right. Viking runic graffiti in Maeshowe, Orkney.

Right. Emigrants leaving the Isle of Lewis

CANADA.

DOMESTIC SERVANTS FOR ONTARIO.

WANTED, for ONTARIO, DOMESTIC SERVANTS for FIRST-CLASS FARM HOUSES.

TO SAIL IN APRIL.
SITUATIONS GUARANTEED.
ASSISTED PASAGES.

In placing the Girls in Situations every effort will be made to place them as near each other as possible.

IMMEDIATE APPLICATION should be made to MAJOR GOODLIFF, ONTARIO GOVERNMENT REPRESENTATIVE, at

46 POINT STREET, STORNOWAY.

CANADA.

FARM WORKERS FOR ONTARIO

WANTED, for ONTARIO, FARM WORKERS
TO SAIL IN APRIL.
WORK GUARANTEED.

For further particulars apply to MAJOR GOODLIFF, ONTARIO GOVERNMENT REPRESENTATIVE, at

46 POINT STREET STORNOWAY

Immediate Application should be made.

RACE OF GIANTS ARE HEBRIDEANS; 400 COMING HERE

Men of Party Which Sailed Yesterday All Five Feet Nine and Over.

FEAR DEPOPULATION

Sheep Farms and Deer Forests Replace the Island Crofter.

By HENRY SOMERVILLE.
Special Cable to The Star by a Staff Correspondent. Copyright.

London, April 16.—The exodus from Hebrides commenced yesterday when 400 Islesmen with their wives and families boarded the C. P. R. steamer Marloch at Lochboisdale, the solitary port of the Isle of South Uist, to Canada.

This party is going to form the Hebridean colony in Red Deer, Al-

A poster advertising the virtues of emigration

Another poster worrying about the arrival of giants from the Hebrides

An Italian ice cream shop in early 20th-century Edinburgh (© Scottish Life Archive, National Museums of Scotland. Licensor www.scran.ac.uk)

A family of Lithuanian immigrants pose for posterity (Scottish Jewish Archives Centre (www.sjac.org.uk))

Pakistani girls playing in a Glasgow street in the 1960s (© Newsquest (Herald & Times). Licensor www.scran.ac.uk)

DNA being rehydrated on rotating wheels

And yet, the relics of some sort of process of cultural transmission along these shores can still be heard amongst the fast disappearing Gaelic speech community.

Both the Irish and Scots dialects of Gaelic contain very unusual locutions, ways of saying everyday things which exist in only 10 per cent of the world's languages, and not at all on mainland Europe. One good example is the translation of a simple phrase, 'I am a Scotsman'. In Gaelic, it is '*Se Albannach a th' annam* – literally, 'It is a Scotsman that is in me.' This active quality is found in more everyday language. The verb 'to have' is *agam* and it means 'is at me', in the sense of possessing something. These and at least 15 other common and fundamental structural differences are shared in one particular area. Along the North African coast, speakers of Berber, Egyptian Arabic and some of the other Semitic languages (including Maltese, the only Semitic language to be written in the Roman alphabet) use the same forms as a Gaelic-speaking crofter on the Isle of Lewis. Along the shores of the Mediterranean there is a whisper, a sense of a nascent speech community and a set of ways of seeing the world passing that way.

Metal drove movement. As the demand and supply of metal objects mounted through the second millennium BC, trading contacts connected communities regularly over long distances. These may have involved several pairs of hands rather than one merchant sailing directly from Iberia to Ireland. It is thought that copper and gold objects were more prized than practical at first and that elites saw them as the badges of status. In the case of gold, they still do. Both metals could be easily worked but their very pliability was also a weakness. When it was discovered that an alloy of tin and copper was much harder and bronze was invented, trade was forced to widen even more.

The only substantial and accessible deposits of tin in Western Europe are found in Cornwall and these pulled mercantile focus northwards to Britain. Pytheas may have known of Britain and been curious because of the tin trade. As merchants plied up and down the Atlantic-facing coasts, language seems to have followed and facilitated their transactions. In a series of books and lectures, the great prehistorian of Europe, Sir Barry Cunliffe, has offered a compelling thesis that Celtic languages developed as a lingua franca along the coasts of Iberia, France, Brittany, Britain and Ireland. As navigable rivers took the trade in metal goods and other items inland, the language was adopted by communities far from the Atlantic coasts.

The distribution of different versions of Celtic languages in the first millennium BC certainly seems to chime with this conjecture. Celtiberian and Lusitanian were spoken in what is now Portugal and Spain, including Galicia, while Gaulish was the speech of all of the area of modern France as well as Rhenish Germany. A dialect known as Lepontic after the kindred who spoke it in the upper Po Valley persuaded the Romans to talk of 'Cisalpine Gaul', literally Gaul on this side of the Alps. British was the speech of virtually the whole island of Britain, the only likely exception being Argyll in south-western Scotland. And Irish was spoken in Ireland.

The dating of language adoption is problematic but the use of clearly Celtic place-names as early as the sixth century BC suggests much earlier beginnings. Elements such as *dunum* for 'fort', *briga* for 'hill', *magos* for 'plain' and *bona* for 'settlement' were in currency at that time throughout much of Western Europe.

All of the Celtic languages are close cousins in the Indo-European family and, if it was originally pulled westwards by the advent of farming 150 generations before, then a framework for understanding and easy adoption will already have existed.

What those pre-Celtic languages sounded like may not be entirely a matter of guesswork or even more scholarly reconstruction. Euskera, the Basque language, has survived against all odds in an Atlantic-facing enclave and it certainly not only predates Celtic but also all the other Indo-European languages. Basic Basque words are very different: *gizon* for 'man', *andere* for 'lady', *neskaro* for 'girl' and *bihotz* for 'heart'. It may well be that the language has survived because its geography prevented outside influence. Distributed on either side of the Pyrenees, the Basque communities live on a rocky Atlantic coastline in what is now Spain and in France behind a string of sandbars and salt marshes known as the Landes. Traders were perhaps reluctant to put in along that littoral and perhaps they did sail diagonally across

Biscay on the open sea, searching for the light burning at the top of the Tower of Hercules.

As sailors plied the busy sea roads between Britain, Ireland and Europe, they will have noticed that the weather was changing – and worsening. More rainstorms swept over the waves and the summers, the time when merchants did most of their business, were growing markedly cooler. And then, in moments and without warning, the mouth of hell opened and spewed forth fire and destruction.

In 1159 BC the Icelandic volcano known as Hekla erupted. Although it was not a supereruption like Toba or Tambora, the explosion took place much closer to Scotland and, at a Volcanic Explosivity Index (VEI) of 5, it had a devastating effect. As a massive tonnage of superfine ash rocketed into the atmosphere, the skies grew dark and those who lived on the Atlantic coasts looked out to the ocean and were fearful. When Hekla erupted in the nineteenth century at a VEI of only 3, the thunder from the mountain was heard in Orkney. In 1159 BC, the roar of the volcano will have boomed across much of north-western Scotland and northern Ireland. No tsunami appears to have left its deadly tidemark but strata of superfine ash have been found at sites along the shorelines. Paleaeobotanists working in the peat bogs of Caithness have detected the dramatic and baleful effects of the eruption. Before 1159 BC, pine trees had contributed 20 per cent of the ancient pollen retrieved in anaerobic core sampling and then it suddenly plummeted to 2 per cent. Analyses of ancient oak timbers retrieved from archaeological sites in Ireland show very narrow tree ring growth for a generation. Hekla brought down a long volcanic winter on the north.

Through the grey and chill mists of dust, the sun could not penetrate, crops did not ripen, only bitter, stunted grass grew and people and animals starved. Prehistorians believe that a long famine gripped the land with perhaps half the population perishing. Almost certainly, migrations to the east began. And, as family bands trudged through the high passes over the Drumalban Mountains, they looked over their shoulders at a landscape that was changing.

More rain had fallen in the four centuries before Hekla and, with lower average temperatures, the weather was remaking north-western Scotland and much of upland Britain. Peat was forming, creeping over the fertile soil and covering farmland under a bitter and acidic blanket. Much of this was not the will of the weather gods but the result of the hands of men.

In the first half of the second millennium BC and long before, there had been a climatic optimum. Average summer temperatures were 38 higher than modern norms. Trees had grown at altitudes of 1,000 feet and communities had lived in what are now wetland areas. But, over a millennium and more, farmers had swung their flint axes and forests had been felled to make way for fields and pasture. The constant need for firewood and a rising population had helped create an open landscape with many fewer trees. When the rain fell and the peat came, the soil was defenceless.

As higher rainfall and lower temperatures combined, organic matter that had died back each autumn did not completely decompose, and as the yellowing bracken and heathers of successive winters piled on top of more and more layers of soft, super-saturated matter, thick strata of peat were formed. It began to suffocate arable land and bury the grass needed to nourish and rear stock. Brown and dank bogs were surrounded by bare, treeless and windswept hills and only a few sheltered straths and coastal grasslands were left less affected.

People had begun to migrate from the west in the second half of the second millennium BC, but after Hekla blew itself apart a migration probably turned into an exodus. There were few Refuges on the Atlantic shores and communities were forced to walk eastwards and to the south.

When the winds and driving rains of the volcanic winter caused by Hekla began to clear and the sun shone once more over the Hebrides and the Atlantic coast, farmers began to adapt, to rebuild their lives. But after the famine and the exodus, there must have been room, an opportunity for new settlers.

Argyll is an ancient name. It means 'the Coastlands of the Gael' and once denoted an area far greater than the modern county boundary. Argyll used to stretch from the Mull of Kintyre all the way north to Loch Broom, almost to Cape Wrath. This west-facing seaboard was worst affected by Hekla and, after its long and baleful winter began to warm, it may be that new people came, led by a new elite who sailed across the North Channel. They brought their language with them, an old version of Irish and the ancestor of Scots Gaelic. At the outset of the first millennium BC, not only did the landscape look much as it does now, the people who began to repopulate it and name it spoke a language that would describe the Highlands and Islands for 3,000 years. Out of the dust and smoke of Hekla, the shape of a new Scotland was beginning to emerge.

6

The Iron Men

✹

THE INFLUENTIAL and quietly perceptive French historian Fernand Braudel proposed that the prehistoric past could be understood better by bearing in mind what he called the *longue durée*. Over tremendous spans of time, he believed, people in similar geographical, social and economic circumstances would retain some similar habits of mind. This allowed prehistorians to make better sense of the deep past by looking at more recent, better documented communities of, say, hunter-gatherers in Southern Africa, New Guinea, the Australian outback or even the Hebridean islands and St Kilda and thereby make links between them and the peoples of prehistoric Europe.

Braudel's approach can be tremendously useful when dealing with practicalities such as land management, seamanship and even political structures. But it should always be borne in mind that our ancestors were, in centrally important ways, very different indeed from us – as L. P. Hartley observed in his novel *The Go-Between*, 'The past is a foreign country: they do things differently there.' In the sand dunes of South Uist in the Outer Hebrides, archaeologists discovered how differently.

The Uists and Benbecula are a long string of islands and islets linked by causeways across a watery landscape. Their geography seems at first disobliging. On the Atlantic-facing western shore where shelter is much needed, the terrain is flat and sand-blown while, in the east, overlooking the calmer inshore waters of the Minch, a screen of majestic, rocky hills rises. But the flatlands are beautiful – most of the coastline is fringed by long, deserted beaches and immediately inland stretches one of the loveliest landscapes in Britain.

The machair is unique to north-western Scotland and in summer it sparkles with colour as thousands of flowers bloom. For about a mile beyond the beaches of the Uists, there is a lush grassland that grows on shell sand, a protein-rich base formed from the crushed particles of millions of seashells. Dredged up from the bed of the Atlantic by a thousand storms and driven onshore, the shell sand became a fertile haven.

As the interiors of the Hebridean islands were blanketed by sterile peat bogs and heather in the second millennium BC, people came to live on the machair. Despite the fact that the ocean could unmake their dwellings in a single stormy night, the dunes shifting in hours to inundate fields and destroy their crops, there is a continuous history of human settlement from 2200 BC to AD 1300. The sand may have been a constant threat to those generations of farmers and fishermen, but to the archaeologists who wished to chronicle their vanished lives it turned out to be a friend.

At Cladh Hallan, near Daliburgh in South Uist, there is a modern cemetery enclosed by a dyke and overlooking the ocean. It is a beautiful, elegiac place to mourn the dead, with the eternity of the mighty Atlantic washing onshore only a few yards away and the sun setting over an immense horizon. Cladh is the Gaelic word for a grave and, although no one realised it, its use in that place was very ancient.

In 2001, a team of archaeologists from Sheffield University were excavating a series of roundhouses built some time around 1100 BC, two generations after the thunder of Hekla had stilled. Six or perhaps seven dwellings had been arranged like a terrace and their foundations – and what lay below them – were well preserved by the windblown sand of the machair.

Concentrating on the most northerly houses, the excavators came across something remarkable and utterly unique in all of British history. In a pit dug under the floor of the north-east quarter of one of the roundhouses, the skeleton of a man was found. With his knees tucked up tight under his chin, his head inclined and his arms folded over, he had been fitted into as small a space as possible. And then soon afterwards another skeleton was discovered, this time that of a woman buried in the southern half of the house. Finally, the archaeologists revealed a third skeleton – that of a baby. These burials appeared to be part of some unknowable foundation ritual performed by the house builders. But, once the bodies had been removed and examined scientifically, more mystery began to swirl around the machair at Cladh Hallan.

It became clear that the skeleton of the man was a composite, the torso and limbs from one individual, the skull and neck from a second and the lower jaw of a third man. Even more remarkably, the torso was found to be 400 years older than the house on the machair. It was built around 1,100 BC but the man buried under the floor had died about 1,500 BC. The woman and the baby were also older, having died around 1,300 BC. What intrigued the archaeologists even more was the fact that the bodies were not true skeletons – enough soft tissue and sinew had survived to indicate that they had been mummified.

Nothing like this had been found before. It was the earliest example of body preservation yet found in Europe. The Cladh Hallan mummies were older than the most famous found in Egypt, those of the pharaohs Seti I and Rameses II (although the earliest dated to 6,000 BC). The obvious question was how. How had the people living on the windy shores of Atlantic Scotland preserved the bodies of their ancestors? Further tests confirmed what the excavators had suspected. Very soon after death, the corpses had been buried in peat bogs. The cold, anaerobic nature of these prevented oxygen from stimulating bacteria to corrupt the soft tissue. Peat may have destroyed the agricultural landscape of the second millennium BC but it seems to have preserved the remains of prehistoric farmers.

From a generalised overview of Britain's prehistoric monuments, it is reasonable to conclude that communities revered their ancestors. Not only were their buried remains an obvious statement of continuity, of the ownership of land by inherited title, they appear to have been worshipped in many cases. At Cladh Hallan those who mummified the bodies of these preeminent individuals took reverence or worship a stage further.

To most of the early tomb builders who interred skeletal remains, the piles of bones and skulls seem to have had a collective importance. And adding to them added to that power. But to succeeding generations one skull will have looked much like another and individual identity was quickly submerged. When single burials became more common in the second millennium BC, the focus shifted. And when the corpses of the recently dead were placed in peat bogs at Cladh Hallan to be preserved, those who retrieved them a few months later will still have been able to recognise their faces. Their names and identities will have lived on. And perhaps they became sacred in themselves, like corporeal effigies, prehistoric versions of plaster saints. Excavators believe that there may have been an ancestral tomb at Cladh Hallan, a house of the dead where the mummies were kept and perhaps worshipped.

By 1,100 BC, something had changed. It may be that the community at Cladh Hallan decided that the mummies (there were at least three men as well as the woman and the baby) had become so decrepit and disarticulated that they should be buried under their houses. That may indeed be why the composite was made – because other pieces of the three male mummies were discarded. In that way, their sacred presence would endure, be part of everyday life. Or new people may have arrived to settle on the machair on South Uist. If Hekla had indeed been tremendously destructive and depopulated the landscape, then the Uists would have been worst hit. Perhaps the new people also revered their ancestors but treated their remains in a different way. Perhaps the techniques of mummification had been lost. In one of the roundhouses, the fresh corpse of a young girl was buried under the floor.

The excavations at Cladh Hallan threw up less sensational but nevertheless valuable revelations. By the outset of the first millennium BC, roundhouses had become a favoured form of domestic architecture all over Britain. Roof-bearing beams were secured on top of circular stone wall heads and tied together at the apex to make a conical shape. Turf, thatch, reeds or brackens were laid on and, while several variants developed, the basic form of the round-house remained in use until the early historic period.

What Cladh Hallan confirmed was the way in which the interior was organised. So that the rising sun could penetrate the windowless building, the door was always set to the east. In the centre burned the fire on its hearth, a patchwork of paving stones encircled by a kerb to keep the embers manageable and where cooking vessels could sit. Food seems to have been prepared in the south-eastern quarter of the house and domestic work done in the south-west. The north was where the occupants slept. There is a suggestion that there may also have been a gender division with females to the south and males to the north. At Cladh Hallan the mummified woman was buried in the southern part of the house.

The prehistoric way of death in South Uist and in other parts of Britain is fascinating, both revealing and mysterious at the same time. But all of the archaeology and scientific testing leaves one central question unanswered. Most of the burials discovered to date, the Amesbury Archer, the people interred in Maeshowe, the young man at Balfarg are all of elite figures, an aristocracy/theocracy of some kind. There are very few of these – and the central question is simple. Where is everyone else? Where are all the other bodies buried?

Even if the population of Britain between 9,000 BC and AD 1 was small, that still leaves millions of corpses unaccounted for. No large prehistoric cemeteries have been found, no mass graves. How did communities and families dispose of their dead bodies? Cremation would account for a complete disappearance since it leaves only dust behind but the tremendous heat required to incinerate a human body would surely have consumed far too much precious wood. That leaves only one viable alternative. Were bodies left exposed to be defleshed by carrion eaters like crows and some other raptors? And were their skeletons then deposited in lochs and rivers? There is some evidence that the Thames was seen as a sacred river and many ancient skulls have been dredged up from its murky silt. But the question hangs in the historical air – where have all our ancestors gone?

Watery places and major rivers like the Thames appear to have become sacred from the middle of the second millennium BC onwards. Perhaps this altered religious focus stemmed from the status of water as a graveyard. Perhaps it was a reaction to a wetter and cooler climate. In any event, it became clear that the prehistoric peoples of Scotland believed that their gods could be revered by the sides of lochs, bogs, rivers and wells. In southern Scotland in particular, ceremonies of sacrifice were held, probably led by a priestly figure on a jetty or causeway built out into the deepest part of a loch. What he did was simple – he threw the most valuable portable items owned by communities into the sacred water. These were metal objects, first made of copper, then bronze and later iron.

Small but dark lochans appear to have been favoured. At Carlingwark Loch in Castle Douglas, Duddingston Loch in Edinburgh, Blackburn Mill in Berwickshire and Eckford near Kelso, large quantities of prehistoric metalwork have been retrieved from the silt and the mud. The find-spots are often too far from the loch shore for the objects to have been thrown and a wooden jetty or perhaps a boat must have been used. At Drinkstone Heights, between Selkirk and Hawick, a small loch was drained in the nineteenth century and a causeway built out into its centre was revealed.

In order for sacrifice to be effective, what is offered has to be valuable and metalwork undoubtedly was. Most objects appear to have been ritually damaged, weapons especially, before being deposited. Perhaps the idea was to avoid any idea of hostility and offer a clear impression of submission. The pantheon of gods identified in later sources for Britain in the latter part of the first millennium BC and into the historic period is very large and apparently chaotic. But one principle does seem clear. These were not gods of love and forgiveness. On behalf of their people, priestly figures hoped to propitiate their gods. That is, they offered valuable metalwork as a sacrifice to persuade the pantheon not to be hostile, not to send rainstorms during the harvest, not to send sickness amongst the flocks and herds and generally not to bring down misfortune on the community.

It seems that those malign and vengeful gods were sought in

the sky as well as the murky deeps. From the beginning of the first millennium BC, what are known as hill forts began to dominate the landscape. The largest in Scotland is also one of the most commanding. From Eildon Hill North, above Melrose, virtually the whole plain of the lower Tweed can be seen, almost as far as Berwick and the North Sea shore. The sheltering Cheviot Hills to the south and the Lammermuirs to the north enclose an immense and very striking vista. There are three Eildon Hills, the place the Romans later named Trimontium, and they rise abruptly out of the riverine landscape.

Some time around 1,000 BC, a huge ditch was dug around the summit of Eildon Hill North. A tremendous undertaking, involving many hundreds of labourers, it measured more than a mile. Using mattocks and baskets, the diggers piled the upcast from the ditch on its inside edge to form a rampart. Stockades of wooden stakes were often driven in to this bank and, for Eildon Hill North, the trimmed trunks of more than 10,000 trees would need to have been dragged up the steep hill. No fewer than five gateways were built and the long perimeter enclosed an area of more than 50 acres.

On a shelf-like plateau to the south of the summit of Eildon Hill North, the remains of 300 footings for small roundhouses have been counted. They show up particularly well after a dusting of light snow. Adding in a scatter of other house platforms elsewhere on the hill, scholars reckon that the hill fort could have accommodated a population of 3,000 to 6,000 – more than live in the modern town of Melrose at its foot.

What was this place? If Eildon Hill North was a permanent settlement, a prehistoric town, then it faced several severe logistical difficulties. Quite apart from the need for food and virtually every other essential domestic item, such as firewood, having to be lugged up a steep hill, there is no spring on the plateau or the summit. Rainwater could have been collected in simple cisterns but what was to be done in a dry summer?

If this complex was intended as a hill fort, a defended site, then the builders created a place impossible to defend. The mile-long perimeter will have required a huge force of soldiers to man it and repel attack, and the construction of five gateways was a military nonsense. Gates in forts or castles were always the weakest points and one would have been sufficient.

Neither a town nor a fortress, Eildon Hill North had to have another very important purpose. It took substantial power and political will to build something as impressive on this difficult site and it may well be that its purpose was exactly that – a clear, commanding and unarguable display of power or a statement in the landscape, as historians have put it. Allied to this, Eildon Hill North seems to have been a sacred site, a role almost certainly indivisible from the political. It is likely that it was a sky temple, the summit consecrated as holy ground. By 1,000 BC, the idea of a sacred precinct marked out from the secular world was already old. Henges such as Balfarg began life as simple precincts, places into which special people processed to perform rituals and somehow commune with the gods. Eildon Hill North is best seen in this tradition, a monumental example of a sanctum sanctorum, a holy of holies.

With a capacity of more than 3,000 souls, this sacred precinct was clearly not seen as exclusive. A whole community or a large part of it seems to have been involved. But what perhaps best explains the phenomenon of Eildon Hill North and the many other massive hill forts built in Britain at that time requires a substantial preliminary.

All over Western Europe, Rome left an indelible cultural legacy. Its military defeat may have been ultimately total but the institutions of the Empire eventually triumphed and principal amongst these was the widespread adoption of the regional and colloquial versions of Latin. The barbarian kindreds who invaded France lost their own languages and began to speak in French. The same surprising transfer took place in Italy and the Iberian Peninsula.

Britain was different. During the four centuries of the province of Britannia, Latin was the language of the army, the government, learning and, later, the church. But it did not replace the dialects of Old Welsh spoken by the native British. While Gaulish and Celtiberian were abandoned in favour of the versions of Latin that would become French, Spanish and Portuguese, no such transition took place in Britannia. And no satisfactory historical explanation has yet been advanced for this.

When the Angles, the Saxons and the Jutes began to arrive in numbers on the coasts of England in the fifth century, they quickly seized political control over wide areas of the south-east. How small numbers of men who sailed from their homelands in Denmark and northern Germany in small boats succeeded so quickly and decisively is also difficult to explain. But succeed they did and, by the seventh century, England was theirs.

Uniquely, the Angles, Saxons and Jutes not only retained their dialects of Low German, what became Early English, they forced the native population to adopt them. This sequence of events happened nowhere else in the old western provinces of the Roman Empire. DNA sampling is inconclusive but it certainly does not suggest a mass influx of settlers to swamp incumbent populations. This did happen on mainland Europe. When the Rhine froze in 406 and a vast horde of Vandals, Suevi and Alans poured into northern Gaul, crossing the river near Coblenz, a true invasion did take place and the DNA of France was altered at that moment. But, in Britannia, it looks very much as though elite war bands gained military mastery and, from that position of brute strength, a small group forced their language on those they had mastered.

In the face of this advance, Celtic languages fled to the west to survive in Cornwall, Wales, Man, Cumbria, eventually north-west Scotland and the west of Ireland where they are still spoken by a tiny remnant. Clinging to the rocky shores of the Atlantic, Celtic speech communities have not only preserved their ways of describing the world but they have also maintained a clear continuity over millennia. Scholars generally accept that versions of Welsh and Gaelic were spoken in Britain and Ireland at least as early as 1,000 BC. That means a cultural as well as linguistic continuity over three millennia – the survival of languages which have described the same places for all that long time. If the principles of Fernand Braudel's *longue dureé* are considered in parallel, it may well be that Gaelic and Welsh carry echoes of our deep and shared past.

And that may have something to say about the tremendous rampart dug around the summit of Eildon Hill North. Above all, Celtic languages describe a rural Britain, a world of farmers, shepherds, animals and cultivation. Even now there are no cities where Gaelic is the language of business and while legislation has recently forced Welsh into certain institutions, it too described the landscape. An old lady living near Badachro in Wester Ross, one of the very last monoglot speakers of Gaelic, used to talk of English as 'the commercial language'. A similar relationship existed in the days of the Roman Empire. Latin was spoken in the towns of Britannia while Old Welsh was heard in the hills and valleys of the province. When Christianity sent down roots in the third and fourth centuries AD, it flourished at first in towns and the words peasant and pagan are cognate – *paganus* simply used to mean a countryman.

Literacy was also characteristic of the cities of Britannia and the administration of government and the army. And it was widespread, reaching down into the lower ranks of a very stratified society. One of the most striking aspects of the letters and lists discovered at the Roman fort of Vindolanda near Hadrian's Wall is that they were usually written by scribes on behalf of both aristocrats and common auxiliary soldiers. Both parties are likely to have been able to read what was written for them, especially when lists were involved. Some of these documents spoke of high society (such as it was on the windy northern frontier) – birthday invitations and hunting parties – but many were simply requisition dockets or lists – orders for food, kit, cart parts and other everyday items. Because these were sourced in the local economy, which must have flourished around the Wall and its huge garrison, it is very likely that local producers could speak Latin and some may well have been able to read and even write it. Illiteracy would not have been allowed to become an impediment to profit.

This cultural collision is the site of another conundrum. While native farmers and tradesman probably had some proficiency in Latin and understood what literacy was, and perhaps its advantages, they never thought to use it for their own language. It is not that there are very few inscriptions or written records of any kind in Old Welsh: there are none. Nothing, it appears, of the language spoken all over Britannia was ever written down over the four centuries of Roman occupation, or indeed before then. Or at least nothing that has survived. It is a puzzling historical blank. Certainly, Latin was exceptional in Europe as such a thoroughly literary language but where others came in contact with it, those non-literate cultures began to adopt its alphabet and write things down, leaving some sort of record.

In Britannia, this exclusion may have involved a choice. The beauty of spoken and sung Gaelic – and there can be few languages more lyrical and expressive – has prompted its admirers to call it the language of Eden. There may be a whisper of history here. Celtic speech may have been the speech of the gods – and it was certainly the language celebrants used to talk to them. To record it in writing may have been sacrilegious. The Druidic priesthood undoubtedly disapproved of any concrete record of ceremonies or liturgy and they committed prodigious amounts to memory. It may be that they feared that the mysteries of what took place inside the sacred precincts would be laid open and revealed to those who did not believe. When Druidic culture was all but destroyed by the XX Legion who attacked the Sacred Isle, Ynys Mon or Anglesey, in AD 60 under the command of Suetonius Paulinus, all of that hidden knowledge was effaced.

What did not belong exclusively in the memories of the priesthood and what huge precincts like Eildon Hill North were intended to accommodate were the celebrations of the turning moments of the agricultural year. These have been preserved in traditions of immense antiquity, in the dictionaries of Irish and Scots Gaelic and Welsh and in habits of mind that have survived in rural Britain well into the historic period.

Each year, four festivals were celebrated, becoming disguised and Christianised as the quarter days. Very ancient, reaching back into the first millennium BC, these important dates were recorded in the Coligny Calendar. This was discovered near Ain in France and it survived because the days and months of the Celtic year had been engraved on a bronze tablet. Profoundly influenced by the phases of the moon, the Coligny Calendar divided each month of 29 or 30 days into a dark half and a light half. In a rural landscape entirely devoid of any artificial light and with cooking fires hidden from sight inside roundhouses, the waxing moon on a clear sky was the only illumination. After sunset in the dark half of the month, the world out of doors was shrouded in blackness, a place where invisible danger lurked.

On the bronze tablet, the year was similarly divided. The dark half began and the light half ended with the festival of Samhuinn. It was to mark this turning point that the roundhouses on Eildon Hill North were reoccupied. It was not a prehistoric town or a military fort but a sacred place where the peoples of the upper Tweed valley came to celebrate the passing of each of the four seasons of the year. Samhuinn is a Gaelic word meaning summer's end and, around the modern calendar date of 31 October, our ancestors climbed up to holy ground like Eildon Hill North to feast, worship, dance, light great bonfires and listen to the words of their priests and kings.

On Samhuinn Eve, the Celtic peoples of Britain believed that the dead drew close to them. In the growing darkness of approaching winter, the veil between worlds wore gossamer thin and, in the darkness beyond the light of the bonfires, the dead walked in the world once more. In the time when the year was dying, when the leaves had fallen and animals were slaughtered, the whispers of ancestors swirled around the sparks and crackle of the flames. Samhuinn survived the coming of Christianity and its traditions lived on in Halloween. For many centuries, the dead were seen in the dark streets of the towns, villages and hamlets of Scotland as young men became guizers. Blackening their faces with soot and dressing in white shrouds, they terrified children and knocked insistently on the doors of the living to ask for food and drink. They carried the samhnag, a lantern kindled only on Samhuinn Eve, and, up on Eildon Hill North and on hundreds of other holy and high places all over Britain, it was used to keep the evil dead at bay. Not all ancestors were benign and the samhnag would ward off the unwelcome. After the eighteenth century, country people began to hollow out turnips and, after carving slits for eyes, nose and mouth, they would fix a candle inside. The turnip lanterns of Halloween are an evocative and direct memory of Samhuinn Eve.

When the peoples of the Tweed valley walked up the steep slopes of Eildon Hill North and swept out their dusty roundhouses and made them snug, they looked up to the summit. There a great bonfire waited. In the first millennium BC in Ireland, men stood by with their torches on the Hill of Tara, looking for a signal. When they plunged them into a huge bonfire and the flames leapt into the night air, they also lit a beacon. On nearby hilltops, inside sacred precincts, others waited to see Tara's fire and, as they lit their own, the celebrations of Samhuinn Eve crackled out across Ireland.

The largest hill fort closest to Eildon Hill North is Yeavering Bell, almost 25 miles due east. On the very summit of the Bell, inside the ramparts, through a nick in the surrounding screen of the Cheviot Hills, it is possible to see Eildon Hill North clearly. Due south rises the conical peak of Ruberslaw, where another large precinct encloses seven acres and, when the fire on Eildon blazed, it would have been seen moments later from that summit. In the opposite direction stands another singular, dramatic hill. On Earlston Black Hill, the ditching of three concentric ramparts can be made out and, while no archaeology exists to confirm their date, they appear to be contemporary with Eildon Hill North.

The place-name of Earlston, the large village at the foot of the hill, has been unusually well documented. It was attached in the Middle Ages to one of the most famous Scots in history, Thomas the Rhymer, Thomas of Ercildoune. His reputation as a prophet was Europe-wide and his name was very old. Ercildoune originally referred to the hill not the village and the second element – from *dunum*, 'a fortified place' – makes that clear. The first part is very intriguing. Not a personal name, as has often been assumed, it may derive from a Celtic root word meaning a meeting place – Dun Airchill is the old word order, Airchill Dun the later.

Both Irish and Gaulish sources talk of the Samhuinn celebrations lasting three nights and involving large assemblies of people. At Tara, they came to hear their kings or their representatives speak the law, their priests offer propitiatory sacrifices and almost certainly to pay their taxes. In Ireland, Samhuinn evolved into the principal festival of the year while, in Gaul, the quarter day of Lughnasa became pre-eminent. Noted in the Coligny Calendar, it was held on the day equivalent to 1 August and was the occasion of an important assembly of Gaulish kings and leaders. This *consilium* was recognised by the Emperor Augustus and, after 12 BC, it began to convene at Lyons, anciently Lugdunum, the fortress of Lugh, the Celtic god who almost certainly presided at Lughnasa.

The political need for an assembly, a large gathering who would listen simultaneously to what kings and their priests had to say, was what caused the mile-long rampart to be dug around Eildon Hill North. In an age before written communication, the law was spoken out loud and clearly to many. Such gatherings reduced the possibility of ambiguity and, when the king or his proxy spoke, there could be no doubt about the fount of justice and decision making. This custom is still enacted on the Isle of Man at the mid-summer meeting of Tynwald, the island's venerable parliament. At midsummer, it meets out of doors at the three-tiered Tynwald Hill at St John's in the middle of Man. While the King of Man (now a representative of the House of Windsor) sat at the top surrounded by his leading men, the Deemsters spoke the law. They still do. At the foot of the Tynwald Hill two bewigged and begowned lawyers, known as Deemsters or Doomsters, recite a list of new laws each year in Manx Gaelic. Called 'breast law', something known by heart, new legislation cannot be enacted until after it has been spoken out loud to the assembled people of Man.

After the fires of Samhuinn had died and those who had gathered around them had descended Eildon Hill North, the winter's dark set in. The earth had begun to grow cold and the dark half of the year waited to be endured. Rural peoples were acutely sensitive to the seasons, they had habits of mind that have almost, but not quite, died out. Here is an extract from an interview of the 1980s with Canon Angus MacQueen, a Catholic priest who lived most of his life on South Uist:

> One of the things that is wrong with young people today is that they don't have as much sense of the life of the earth as we had. When I was a boy I used to take off my winter boots on the day of the Feast of St Bride in March and not put them back on until the Feast of All Saints [Samhuinn] at the end of October. Walking over the machair to school I could feel the earth coming alive through the soles of my feet. And in the autumn I could feel it getting ready to die again.

Winters could be very hard but, for those prudent enough to have stored grain and cured meat, the frosts and snows will have been tolerable. The first signs that the earth was coming alive again could be seen in early February. At the feast of Imbolc, the peoples of the Tweed valley climbed Eildon Hill North once more. As they shivered in the bitter winds of the late winter and fires were lit on the summit, they celebrated the stirrings of life. Imbolc means 'in the belly' and refers to pregnant ewes. By the beginning of February, these were lactating and could be milked before giving birth and fresh cheese could be made.

As Canon Angus knew well, the festival and its goddess had been Christianised as St Bride's Day. Associated with fire, purification, fertility and high places, Brigid was said to walk the earth at Imbolc and offerings were left out for her. Worshippers asked the goddess for an early end to winter and a very old Gaelic verse remembers not only her but also the emergence of hibernating animals like badgers and snakes:

> Thig an nathair as an toll
> La donn Bride
> Ged robh tri troighean dhen t'sneachd
> Air leac an lair.

> The serpent will come out of the hole
> On the brown day of Brigid
> Though there should be three feet of snow
> On the flat surface of the ground.

By the time fires blazed on hilltops at Beltainn, the earth had come alive, and in the early days of May the light half of the year at last began. It was welcomed by blossom, a joy recalled in the old English phrase, 'Here we go gathering nuts in May.' There are no nuts in May, it being far too early in the year, and the reference is to knots of greenery and blossom. In Padstow, a fishing village on the north coast of Cornwall, the streets are decorated on Beltainn Eve with boughs of the May Tree and hats and clothes with cowslips, the first flowers of the year. This is a floral prelude to the May Day rites of the 'Obby 'Oss, a celebration of fertility from a time beyond memory.

In Padstow (and in other south-western villages up until the nineteenth century), the 'Obby 'Oss dances around the garlanded streets. Much drink is taken in what feels like an authentically Celtic festival. Inside a circular hooped frame covered with canvas and topped with a strange carved head that looks more African than Cornish and nothing like a horse, the 'Oss dances to the music of time. Undoubtedly a stallion, he pulls young women under the skirts of his costume and, accompanied by whoops and shrieks, he behaves a little like a stallion. Through the crowded streets and lanes, the 'Oss's supporters imitate the squeals of mares in season when they shout, ' 'Oss, 'Oss! Wee 'Oss!' It is a remarkable survival.

The dance of Padstow's stallion is well timed. Given a gestation period of 11 months, the optimum time to put mares in foal is May. Spring grass the following year will make for good, nourishing milk and better weather for delicate newborns. Horses were central to the Celtic culture of the first millennium BC. Horse harness begins to be found in some quantity (at least its metal parts do) after 1,000 BC and under the sacred rampart around Eildon Hill North, a sacrificed pony was buried near a gateway.

As well as equine fertility, the period around Beltainn was the prompt for the journeys of transhumance. After the thaw and spring rains, green shoots had begun to poke through the brown dieback on upland pastures. If ewes had lambed in March, as most do now, then their offspring would have been strong enough to walk up the hill trails but still sufficiently attached to their mothers to stay with the flock on what could be a dangerous few days and nights. The chief role of early pastoral dogs seems to have been to bark and act as a warning of wolf packs circling.

Lugh, sometimes called Little, Stooping Lugh, was one of the most idiosyncratic of the Celtic pantheon. In Irish Gaelic the translation of his extended name was Lugh Chronain, a name eventually anglicised into leprechaun. Revered all over the west, from Gaul to the north of Scotland, he was seen as a trickster, a shape-shifter and generally capricious. More specifically Lugh was associated with that most magical of skills, metalworking. The popularity of his cult saw his name left on the names of European cities – not only Lyons but also Leyden, Laon and, more uncertainly, London.

In the A to Z of the latter, Ludgate Hill is definitely a memory of Lugh, the leprechaun.

The quarter-day festival held at the beginning of August is sometimes seen as a harvest celebration but the dates are too early, especially in the north. More likely it was a time when the wild harvest of fruits, berries and nuts could be ingathered and enjoyed, the first bounty of the year. Lughnasa was also a time for what were known as handfastings. A pre-Christian tradition, this was, in essence, a form of trial marriage when a couple contracted to cohabit for a year and attempted to conceive children. In a rural economy, the fertility of human beings was as important as that of animals for many hands were needed to work the land, and if a couple found that they were barren it presented a profound economic difficulty. Handfasts could be dissolved by either party after a year, and if any children were produced by a couple who decided they were incompatible these offspring were cared for by an extended family. Despite the evident disapproval of the church, trial marriages of this sort were still being arranged in the Scottish countryside as late as the nineteenth century and were also recognised legally until 1939 when the Scottish marriage laws were reformed.

On a commanding eminence overlooking the confluence of the Black Esk and the White Esk in Dumfriesshire, the ditches and banks of a large hill fort can clearly be seen despite the dense forest crowding around. Castle O'er is a remote place now, lying on the southern edge of the sprawling Craik Forest but in 1,000 BC many people lived here. The Ordnance Survey for the area is speckled with prehistoric sites. Before the trees were planted, this was a productive region where many sheep were run on upland pasture and free-draining meadows. When the traveller and historian Thomas Pennant came to Castle O'er in 1772, he saw a landscape not much changed over two millennia – and he witnessed customs that had also altered little:

Among the various customs now obsolete, the most curious was that of handfisting [sic], in use about a century past. In the upper part of Eskdale an annual fair was held where multitudes of each sex repaired. The unmarried looked out for mates, made their engagement by joining hands, or by handfisting, went off in pairs, cohabited till the next annual return of the fair, appeared there again and then were at liberty to declare their approbation or dislike of each other. If each party continued constant, the handfisting was renewed for life but, if either party dissented, the engagement was void and both were at full liberty to make a new choice.

The language of love – or, at least, contract – in Britain and Ireland in the first millennium BC was not uniform. Terms of endearment varied a great deal. Underlying an inevitable pattern of regional dialects, there was a fundamental divide in the Celtic speech of the Isles. This difference can be clearly heard today. In the Gaelic-speaking districts of Scotland and Ireland known as the Gàidhealtachda, people use the version labelled by linguists as Q-Celtic. When they use, for example, the words for 'head', 'four' or 'son of', they say *ceann*, *ceithir* and *mac*. In Wales, Brittany and amongst the enthusiastic revivers of Cornish, they will say *pen*, *pedwar* and *mab* or slight variants. These differences define them as speakers of P-Celtic.

It appears that the latter came later. Q-Celtic is the oldest language still spoken in Britain and Ireland, reaching back perhaps into the second millennium BC. Around the seventh or eighth centuries BC, a language shift occurred and what is now England, Wales and most of Scotland began to adopt P-Celtic. The dynamics of this shift are fascinating and are made more patent by a study of DNA markers and their movement and, as time goes on, by the observations of people who came to the Isles from the Mediterranean.

Herodotus never made the journey across Europe and, when the Father of History did write about Britain, he was characteristically cautious, heavily qualifying a few scant remarks. The word history derives from a Greek word meaning 'a witness' and this was the sense in which Herodotus saw himself as a historian. Born in Halicarnassus (the modern Turkish city of Bodrum) some time early in the fifth century BC, he did indeed witness – or knew people who had witnessed – world-changing events. It was these that prompted him to write.

Opening with the legalistic phrase, 'Here are presented the

results of enquiries carried out by Herodotus of Halicarnassus', he sounds more like a policeman in court. He immediately establishes his grand theme – the wars between the Persian Empire and the Greek city states. In 490 BC, the army of the Athenians and the Plataeans defeated a much larger invasion force at Marathon and, when Xerxes led the Persians levies into Greece once more in 481bc, they were again repulsed. Herodotus saw himself as the chronicler of the great – and astonishing – turns of history and they form the central theme of his sprawling *Histories*.

The battle on the plain of Marathon does not appear until Book Six and a vast mass of introductory material contains much that is colourful, amazing and puzzling. Herodotus had heard of the Tin Islands and thought they probably lay somewhere far away in the Northern Sea. 'But,' he wrote, 'I cannot speak with any certainty.' More sure about the location of the peoples known as the Celts but wildly wrong about the source of the Danube, the river he calls the Ister, Herodotus noted:

> The Ister rises in the land of the Celts, at the city of Pyrene (the Celts live beyond the Pillars of Hercules) and are neighbours of the Cynesians who are the westernmost European people, and flows through the middle of Europe . . .

This, the earliest record of any kind of the location of Celtic peoples in the Iberian Peninsula, is confirmed by later sources. The Pillars of Hercules are of course the Straits of Gibraltar but who the Cynesians were is lost to history. Their neighbours, the Celtiberians and Lusitanians, spoke Q-Celtic and, supported by the evidence of DNA links and the consistent weight of ancient traditions such as *The Book of Invasions* and the Declaration of Arbroath, it seems that that variant of the language developed as a lingua franca as it travelled north to Ireland and Britain by peaceful means. The sons of King Breogán may have climbed the Tower of Hercules, spied the green shores of Ireland on the horizon and boarded their ships to invade but trade is just as effective a means of language transmission as the point of a sword. The commercial links along the Atlantic littoral flourished between 1,300 BC and 600 BC. Goods were fed in and out of the network as they travelled up and downriver from the continental interior. Archaeologists have found ample evidence of a complex system of exchange that operated for seven centuries.

Around the middle of the first millennium, there appears to have been a recession of some kind with trading volumes dropping significantly. Contacts with Ireland ceased and it seems that Britain began to look eastwards. In central Europe, north of the Alps, a rich metalworking culture known as Halstatt after its type site was growing powerful and influential. Its chieftains thrived as their smiths manufactured weaponry and all sorts of other artefacts out of bronze and then iron. Dominated by warriors – or, at any rate, men who prized war gear – much of what is known about the Halstatt peoples comes from the grave goods preserved in their elaborate tombs. Often their chiefs or princes were buried with four-wheeled wagons and, at Hochdorf near Stuttgart, one man's corpse was laid out on a couch made entirely of bronze. An enormously valuable item, it was buried along with weapons, a large cauldron (which probably contained mead), drinking horns and hunting gear and the wagon was laden with dishes and other objects. These were the accoutrements of a wealthy warrior aristocracy. The Halstatt hill forts were impressive and some of the largest commanded river routes and trading trails.

In the second half of the first millennium BC, the power and reach of the chiefs and princes appeared to decline and the focus shifted westwards. La Tène, the name of a later type site, this time a lakeside in Switzerland, began to develop a distinctive style of metalworking that was more identifiably Celtic in character. In the valley of the River Marne in northern France, elite burials changed and included horse harness and two-wheeled chariots as well as the more familiar array of weaponry and drinking kit.

Language historians believe that the Halstatt and La Tène societies spoke a form of Celtic, more specifically P-Celtic. As trade along the Atlantic coasts declined around 600 BC, contacts between Britain and these vigorous power centres increased. It seems likely that, in addition to the fearsomely sharp swords, cauldrons and horse gear, P-Celtic was imported into Britain from the east and it began to overlay the native dialects of Q-Celtic. How mutually intelligible these variants will have been is important and, while modern Gaelic and Welsh speakers do not understand each other, it may be that their ancestors did 2,500 years ago.

A merchants' handbook known as the 'Massaliot Periplus' was complied some time in the sixth century and it offered intelligence and advice to those sailing as far north as Britain and Ireland. The original text of the guide is lost but two place-names it included have survived in later sources. The Periplus talked of Britain as 'Insula Albionum' and the Irish as 'gens Hiernorum'. Remarkably, both names are still part of the modern Gaelic and Irish lexicons. Alba, pronounced 'alapa', is Gaelic for Scotland (and an old name for the whole island of Britain) while Hierne changed little to become Eire and the first element of the English name, Ireland.

Now, both of the names in the sixth century Massaliot Periplus are Q-Celtic but by the time another Greek from Marseilles, Pytheas, came to Britain, one of them had changed. By the late fourth century BC, the *Albiones* had become the *Priteni*. This is a P-Celtic name that means 'the Tattooed People' or 'the Painted People' and it is the origin of the modern name of Britain. Early Welsh talked of its speakers as *Brythoniaid* and, in Q-Celtic speech, the use of Alba shrank northwards finally to apply to only Scotland. *Priteni* in Q-Celtic renders the difference neatly. If P-Celtic had not crossed the Channel and the North Sea from Europe between 600 BC and 300 BC, the British would have been known as the *Cruthen*, the Q-Celtic version of *Priteni*. And the British would now be the *Crutish*.

Archaeology has tracked the northern advance of P-Celtic – dramatically. At the Newbridge Interchange west of Edinburgh, the M9 meets the A8 and leads on to the M8 and Glasgow. As traffic thunders under the bridge or waits at the traffic lights at the roundabout on top of it and aeroplanes drone overhead, leaving or landing at Edinburgh airport, few notice another circle immediately adjacent. The round Bronze-Age burial mound at Huly Hill was made at a time when the landscape was quiet and green. Around it, at least three standing stones were raised and across the roar of the M9 there is a fourth, an outlier almost lost in the midst of an industrial estate. Huly is a Scots word for 'gentle' and, despite the twenty-first century racing past the ancient hill, the place somehow preserves the peace of the long past.

In 2001, planners granted permission for an extension to the Newbridge Interchange. The construction workers were made aware of the proximity of archaeology and consequently they were careful and painstaking. When they came across a pit and what appeared to be remains of some sort, work was halted and experts called in. What they discovered changed perceptions of prehistoric Scotland.

It was a chariot burial. All trace of a body had perished in the soil but the outlines and metal fittings of a two-wheeled chariot were clear. With its dead charioteer probably laid lengthways on top, it had been buried upright with slots dug to accommodate the wheels and the pole. Carbon dating placed the grave in the fifth century BC and, crucially, it made clear links with not only the La Tène culture of west central Europe but also archaeology carried out in south-east Yorkshire.

While the style of the Newbridge inhumation implied direct links with Europe and possibly the valley of the Marne, similar burials, in which the chariots had been taken apart, found in Yorkshire are informative. They were discovered in the territory of a small kindred called the Parisii, a group who appear to have immigrated from the area around Paris. A kindred of the same name lived on the banks of the Seine and gave their name to the city. The Parisii on the bank of the Humber retained their separate character long enough for the provincial government of the province of Britannia to grant them the status of a *civitas* and allow them a tribal capital at Brough on Humber.

In his wonderfully well-written *Agricola*, Tacitus opened the biography of his father-in-law with a general description of Britain. Although maddeningly miserly with names, both of people and places, he offers a fascinating general description of what he and his fellow Romans found. It is worth quoting at length:

Nowhere is the dominance of the sea more extensive. There are many tidal currents, flowing in different directions. They do not merely rise as far as the shoreline and recede again. They flow far inland, wind around, and push themselves among the highlands and the mountains, as if in their own realm.

As to what human beings originally inhabited Britain, whether native-born or immigrants, little has been established, as is usually the case with barbarians. Be this as it may, their physical appearance is varied, which allows conclusions to be drawn. For example, in the case of the inhabitants of Caledonia, their red-gold hair and massive limbs proclaim German origin. As for the Silures, their swarthy features and, in most cases, curly hair and the fact that Spain lies opposite provide evidence that Iberians of old crossed over and settled this territory. Those nearest to the Gauls also resemble that people. Either their common origin still has some effect or, since the two countries converge from opposite directions, shared climatic conditions produce the same physical appearance.

All the same, it is plausible on a general estimate that the Gauls occupied the adjacent island. You can find their rites and their religious beliefs. The language is not much different, likewise the same boldness in seeking out danger – and, when it comes, the same timidity in facing it. Still, the Britons display more ferocity, having not yet been made soft by prolonged peace. We are told, indeed, that the Gauls, as well, used to be warriors of repute. Then decadence set in, hand in hand with peace: their courage has been lost along with their liberty. The same has happened to the Britons long since conquered. The rest are still like the Gauls once were.

Their infantry is their main strength. Some of their peoples also engage in battle with chariots. The nobles are the charioteers, their clients fight for them. In former times the Britons owed obedience to kings. Now they are formed into factional groupings by the leading men. Indeed, there is nothing that helps us more against such very powerful peoples than their lack of unanimity. It is seldom that two or three states unite to repel a common threat. Hence each fights on its own, and all are conquered.

The climate is miserable, with frequent rain and mists. But extreme cold is not found there. The days last longer than in our part of the world, the nights are bright and in the most distant part of Britain so short that you can hardly distinguish between evening and morning twilight. If clouds do not block the view, they say that the sun's glow can be seen by night. It does not set and rise but passes across the horizon. In fact, the flat extremities of the earth, casting a low shadow, do not project the darkness, and night falls below the level of the sky and the stars.

The soil bears crops, apart from the olive and the vine and other natives of warmer climes, and has an abundance of cattle. The crops ripen slowly but shoot up quickly. The cause is the same in both cases, the abundant moisture of land and sky. Britain contains gold and silver and other metals, the booty of victory. The Ocean also produces pearls, but they are dusky and mottled. Some attribute this to the divers' lack of skill, for in the Red Sea the oysters are torn from the rocks alive and breathing, in Britain they are collected as and when the sea casts them up. For myself, I would find it easier to believe that the pearls are lacking in quality than that we are lacking in greed.

By 'Caledonia', Tacitus meant the region of eastern Scotland north of the Tay and at least as far up as the Dee. Later evidence also places the Caledonii in the same general area. The Silures held South Wales and Gaul was of course France. What is striking about Tacitus's observations is that they chime with both the archaeological and genetic stories. Caledonia was where the hall builders of Balbridie settled, it has a clear affinity in its DNA with the Germanic shores of the North Sea and, while red hair is much more common in Scotland than in Germany, the general cultural instincts behind the historian's comments are well founded.

Contact between Britain and Iberia is also recognised – although again the use of hair colour as a link is dubious. Clearly, the amusing description of Britain's climate by a shivering observer more used to Mediterranean sun is somewhat partial. But the description of the long summer days and the long summer nights shows a good deal of Roman knowledge of Scotland and the north. Earlier in his text, Tacitus notes the Roman subjugation (surely nominal) of Orkney and that their fleet circumnavigated Britain.

By the second half of the first century AD, the period dealt with in the *Life of Agricola*, P-Celtic had become the speech of most of Britain, including Caledonia. Gauls and Britons, especially in the south, are likely to have found each other mutually intelligible. The close contact with Rome and its advancing empire led to a good deal of borrowing. While Q-Celtic contains very few words derived from Latin, P-Celtic has many: *mur* for 'wall', *ffenestr* for 'window', *gwydr* for 'glass', *cegin* for 'kitchen', *cyllel* for 'knife', *ffwrn* for 'oven' and *sebon* for 'soap' might all be said to be characteristic of Mediterranean domesticity imported into Britain. Literacy also donated words: *llyfr* for 'book', *ysgryf* for 'essay' and *awdwr* for 'author'. Even *caer* for 'fort' came from *castrum* and *ffos* for 'ditch' is from *fossa*. In all, around 600 commonly used words in Welsh come from Latin. There can be no doubt that the bulk of this substantial transfer took place during the four centuries of the Roman province but the process seems to have been underway some time before, as Rome encountered Celtic speakers in the second half of the first millennium BC.

What this brief review of language history reveals is that the oldest cultural continuity in Britain is also one of the most fragile. It survives in the mouths of the fewer than 50,000 speakers of Scottish Gaelic who use words and habits of mind not unfamiliar to those who lived in the north of Britain at the end of the second millennium BC.

7

The Glory Road

NOT ONLY WAS THE weather miserable, with frequent rains and mists, Britain was also not worth having. The phenomenal expansion of the Roman Empire was driven by what Tacitus called the *pretium victoriae*, the 'wages of victory' or how much wealth could be extracted from the defeated by the conquerors. A sodden landscape, half-hidden by cloud, producing nothing more exciting than cattle, corn and a few substandard pearls, the place was thought simply incapable of delivering a decent return on all that outlay of men, materials and money. Roman commentators dismissed a conquest of Britain as making no sort of economic sense.

But it was good politics. Just as it appeared to Herodotus, Britain seemed to Rome and Romans to be a long way away, far to the cold and barren north, on the edge of the known world. That was the political point. When Julius Caesar courted near disaster with his expeditions in 55 and 54 BC, it was precisely because it was an island at the limits of the world that his daring exploits created a sensation in Rome. Twenty days of feasting and public thanksgiving were ordered by the Senate, thrilled that the Empire could reach to the ends of the Earth and defeat the unimaginable savages who lived there. Glory was the attraction – not profit.

When Claudius was set on the imperial throne by the Praetorian

Guard after the inevitable assassination of the crazy Caligula in AD 41, there was a pressing need for glory. Plots and attempted coups flickered around the court of the new emperor and his grasp on authority seemed to be slackening. Military success would persuade waverers and it was decided that a conquest of Britain would silence criticism. To extend the Empire to the edge of the world? To outdo the great Julius Caesar? Just what Claudius needed. And maybe the pearls were better and maybe there were seams of gold and silver in the misty mountains? An added attraction was that Britain was a good reason for breaking up a potentially dangerous concentration of crack troops stationed on the Rhine frontier. It was a ready-made army for an ambitious general considering a coup. The legions on the German borders would no longer be bored, staring at the impenetrable forests, patrolling beyond the river, searching for phantoms amongst the trees. Britain beckoned.

In the summer of AD 43, led by Aulus Plautius, Rome triumphed in the south of England. When the good news reached the imperial court, the invasion force was commanded to halt on the line of the Thames, consolidate and wait for Claudius. In order to extract full propaganda value, the emperor would be seen at the head of his legions, in his armour, marching with them on the road to glory. To add to the spectacle, he brought elephants, war elephants – creatures never before seen in Britain – and, as the British kings retreated, awed and ultimately defeated, Claudius and the many senators brought along to serve as witnesses entered Camulodunum, modern Colchester. It was the royal centre of the Catuvellauni and their king, Cunobelin, Shakespeare's Cymbeline. In his moment of stage-managed triumph, having gone further than Caesar, Claudius was not to know that, while the south of Britain was won at Colchester, it would almost be lost there only a few short years later.

Eleven kings arrived at Cunobelin's old capital to make submission, according to a triumphal arch erected in Rome, and later the emperor took the title of Britannicus. Amongst those British kings who bowed to Rome was the king of Orkney, his arrival being a cause of yet more amazement at the reach of this new and dynamic emperor. A historian called Eutropius wrote, 'Claudius added to the Empire some islands lying in the Ocean beyond

Britain, which are called the Orkneys.' It was a high-water mark.

Careful not to overextend, the legionary legates marched against only those kingdoms that would not submit. The future emperor, Vespasian, led the II Augusta west into the territory of the Durotriges and laid siege to the vast, sprawling hill fort of Mai Dun, Maiden Castle. The beautifully undulating banks and ditches did provide a defensive shelter as, for once, the sacred precinct took on a military purpose. After batteries of siege engines had bombarded the stockade with stones and huge javelins, the Romans broke in and destroyed the makeshift garrison. Elsewhere, more peaceful methods were deployed. North of Colchester lay the lands of Prasutagus, king of the Iceni, and he had come to terms with Rome, adopting the status of a client kingdom.

By AD 60, much of the south of England had fallen under the control of the invading legions. In his *Annals of Imperial Rome*, Tacitus listed a series of wars against British kings and made much of the story of Caratacus. For nine years this king of the Catuvellauni, the son of Cunobelin, had led a determined resistance. What is striking is the degree of unity and military cooperation across Britain. Even though he had lost his own kingdom (perhaps fleeing with an elite war band) in the early years of the conquest, Caratacus was accepted as a general by those kindreds who refused to submit. This implies a degree of cultural continuity, a shared language and shared beliefs.

Some time around AD 50, Caratacus's army faced the brigaded legions on ground of his choosing in North Wales, probably at Llanymynech Mountain on the modern border with Shropshire. After the discipline and tenacity of Rome's professionals inflicted a surprise defeat, Caratacus fled for sanctuary to Queen Cartimandua of the Brigantes. Tacitus believed this federation of kindreds to be the largest and most powerful polity in Britain. It compassed both sides of the Pennines, modern Yorkshire, Lancashire, much of Cumbria, Durham and Northumberland south of the Tyne. But, instead of using the strength of the Brigantes in the struggle with Rome, Cartimandua had become a client like Prasutagus of the Iceni and she had Caratacus arrested and handed over to the Romans. Paraded through Rome in chains, he was pardoned by the Emperor Claudius but he never saw Britain again.

Resistance continued. In his *Annals*, Tacitus recorded that the kindreds of Wales and the Silures of the south in particular were exceptionally stubborn. And Venutius, the consort of Cartimandua, led a breakaway Brigantian army reinforced by help from outside. It seems that the kings of what is now southern Scotland were being drawn into the fight against Roman domination. Perhaps they came to understand that imperial policy at that time favoured the conquest of the whole island. Certainly what happened AD 60 suggests as much.

Suetonius Paulinus, a new governor, fresh from Rome and no doubt with a clear grasp of the emperor's wishes, led the legions to a corner of Britain that could easily have been bypassed and had little strategic significance. Here is the relevant passage from Tacitus's *Annals* describing the assault on Anglesey:

> So Suetonius planned to attack the island of Mona, which although thickly populated had also given sanctuary to many refugees.
>
> Flat-bottomed boats were built to contend with the shifting shallows, and these took the infantry across. Then came the cavalry; some utilised fords, but in deeper water the men swam beside their horses. The enemy lined the shore in a dense armed mass. Among them were black-robed women with dishevelled hair like Furies, brandishing torches. Close by stood Druids, raising their hands to heaven and screaming dreadful curses.
>
> This weird spectacle awed the Roman soldiers into a sort of paralysis. They stood still – and presented themselves as a target. But then they urged each other (and were urged by the general) not to fear a horde of fanatical women. Onward pressed their standards and they bore down on their opponents, enveloping them in the flames of their own torches. Suetonius garrisoned the conquered island. The groves devoted to Mona's barbarous superstitions he demolished. For it was their religion to drench in the blood of prisoners and consult their gods by means of human entrails.

Even across two millennia, Tacitus's double standards raise an eyebrow. Roman armies regularly slaughtered captives in their thousands or sent them to suffer humiliating deaths in the arenas of their cities for the amusement and titillation of citizens. The Empire shed far more blood for much longer than Druid priests did. Nevertheless human sacrifice appears to have been part of British religious practice both before and after the arrival of the legions.

Across northern Europe, in Ireland, Britain and Scandinavia, more than a hundred bodies dating from 600 BC to AD 400 have been retrieved from peat bogs, preserved by their anaerobic properties. Many show evidence of a ritual death. Perhaps the clearest signs were found on a body discovered in Cheshire known as Lindow Man. After a post-mortem, scientists concluded that he had been an aristocrat. His hands showed no evidence of manual work and the man was young and healthy. Perhaps he was himself a priest or his status in itself recommended him as a sacrificial victim.

Lindow Man's death was protracted and almost certainly excruciating – there was no evidence that he had been given a drug. Almost certainly surrounded by priests and perhaps a large congregation gathered to witness an event of immense significance, the young man was first poisoned and then beaten. He was hit on the head with an axe but the blow did not kill him. He lived to be garrotted and have his throat cut. When the priests placed his naked body in Lindow Moss to drown, it is possible that, even at that moment, he was still alive.

The victim suffered a multiple death, a rite sometimes known as the triple death, and this savagery survived in Druidic traditions well into the Dark Ages. Merlin or Myrddin was said to have been hit on the head, garrotted and drowned in the River Tweed.

Suetonius Paulinus's attack on Mona appears to have been emblematic. Julius Caesar recognised the political power of the Druidic priesthood when he campaigned in Gaul and recorded their use of groves as places of worship and sacrifice, the places destroyed by soldiers on Mona after the battle at the Menai Straits. He also believed that the cult originated in Britain and had been imported into Gaul. Such was the power of Druids that it was said their arrival could stop a battle. And Caesar's account of his campaigns, *Commentarii de Bello Gallico*, was very influential.

In any event Mona, the Sacred Isle, appears to have been the centre of the cult. At Llyn Cerrig Bach, a small lake in the west of the island, metalwork from all over Britain was ritually deposited, probably by pilgrims. Tacitus commented that Mona had given sanctuary to many refugees, displaced people drawn to the island because of its sacred power. There the gods would protect them.

As the legions swept aside the curses of the Druids and the screams of the black-robed Furies with their spiked, limewashed hair and hacked down the groves of the holy sanctuary, trouble was stirring on the other side of Britain.

Two Roman statutes of the second century BC, the *lex Porcia* and the *lex Sempronia*, gave, amongst other things, directions for the public flogging of non-citizens. They were to be stripped naked and bound face first to a post in a place where it was convenient for a crowd to gather and watch. Ancient justice had always to be seen to be done. Whips often had small pieces of metal or bone attached to their tips so that blood would be drawn and the backs or buttocks of prisoners badly flayed. Screams and extreme suffering were also understood as important parts of the judicial spectacle. Flogging was often a prelude to crucifixion and the pattern of punishment meted out to Christ followed the tenets set out in the *lex Porcia* and *lex Sempronia*.

Possibly at Thetford, where the remains of an Icenian royal palace have been found, Queen Boudicca was led to a whipping post, stripped, tied to it and flogged. Her crime had been to object. Boudicca's husband, King Prasutagus, had died and, in his will, bequeathed half of his estates and treasure to the Emperor Nero and half to his wife and two daughters. It was a device often used by Roman aristocrats who hoped that, by giving the emperor a stake in them, the conditions of their wills would be carried out and have the authority of the state to enforce them.

In AD 60, the administration of the province of Britannia was run by the procurator, Decianus Catus, and, while the governor was away campaigning in the Druid sanctuary of Mona, he took matters into his own hands. Catus's initiative was to have catastrophic consequences. Ignoring the terms of Prasutagus's will, he treated the kingdom of the Iceni like conquered territory, looting treasure, taking over estates and evicting aristocrats from their property. When

Queen Boudicca resisted this patent injustice, she was humiliated by a public flogging and her daughters, the royal princesses, were raped by Roman soldiers. It was impossible for Catus to understand that, in Celtic society, women had status and could rule. In Rome, women had the same legal standing as children.

It was an incendiary moment. The kingdom flared into rebellion as the Iceni were joined by the Trinovantes to the south and Boudicca quickly mustered a huge army. She immediately led it south to Colchester. The town was burned to the ground as more war bands rushed to join the rolling momentum of the uprising. When news reached Suetonius Paulinus, the legions hurried southeast and, from Lincoln, part of the IX Legion marched to meet the Queen's growing host. Under the command of Petilius Cerialis, the infantry of the IXth were overwhelmed and slaughtered. With only his cavalry, the tribune fled to take refuge at the fortress at Longthorpe, just to the west of Peterborough. The province of Britannia teetered on the edge of extinction as the thunder of rebellion boomed out over the land.

Meanwhile, Suetonius Paulinus had reached London at the head of an advance party but, when news of the defeat of Cerialis came, a difficult decision had to be made. Opting to regroup with his main force of infantry, still marching down Watling Street from Mona, the governor abandoned London. Tacitus takes up the story:

> Unmoved by lamentations and appeals, Suetonius gave the signal for departure. The inhabitants were allowed to accompany him. But those who stayed because they were women, or old, or attached to the place, were slaughtered by the enemy. Verulamium [St Albans] suffered the same fate.
>
> The natives enjoyed plundering and thought of nothing else. Bypassing forts and garrisons, they made for where loot was richest and protection weakest. Roman and provincial deaths at the places mentioned are estimated at seventy thousand. For the British did not take or sell prisoners, or practise other wartime exchanges. They could not wait to cut throats, hang, burn and crucify – as though avenging, in advance, the retribution that was on its way.
>
> Suetonius collected the fourteenth and detachments of the

twentieth, together with the nearest available auxiliaries – amounting to nearly ten thousand armed men – and decided to attack without further delay.

Stiffened by their habitual discipline and steeled by the certain knowledge that defeat meant annihilation, the legions drove into the vast native army and shattered it. Hemmed in by the wagons of their own camp followers, the huge host was trapped in a blood-soaked killing field. Boudicca and her daughters took poison and, after the battle, Suetonius Paulinus embarked on expeditions of vengeance. The territories of the Iceni and Trinovantes were plundered and burned and those kingdoms that had remained neutral were similarly punished.

What fuelled the uprising (sparked by Boudicca's humiliation) was the threat of slavery and destitution – even in peacetime. Here is Tacitus again:

> They particularly hated the Roman ex-soldiers who had recently established a settlement at Camulodunum [Colchester]. The settlers drove the Trinovantes from their homes and land, and called them prisoners and slaves. The troops encouraged the settlers' outrages, since their own way of behaving was the same – and they looked forward to similar licence for themselves.

Colonisation of this sort was common in the Roman Empire as soldiers completed their terms of service and decided to settle in the provinces. Over the four centuries of Britannia, many veterans were given land, married native women and introduced their DNA. But it is very difficult to identify its traces with any certainty.

By the time of the Boudicca uprising in AD 60, the Roman army was no longer a force of citizens of Italian origin. Augustus, the first and perhaps most clear-sighted emperor, created a highly professional standing army of 28 legions, each with approximately 6,000 men. The number of auxiliary troops was about the same, making a total complement of 336,000. The length of legionary service was extended from six to twenty years. What lay behind Augustus's reforms was, of course, a series of political judgements.

The old republican armies of citizens and assorted levies had developed intense loyalties to their generals, the aristocrats who rewarded them – men like Pompey, Caesar and Mark Anthony. By insisting on an oath of allegiance to the emperor alone, Augustus loosened these bonds and, at the same time, recruitment was widened well beyond Italy to include the provinces. By AD 100 and the reign of Trajan, the ranks of the Roman army were mostly filled by men from northern Europe and only about 40 per cent originated around the shores of the Mediterranean. In AD 200, 80 per cent of the army was from the northern and eastern provinces.

By its ruthless nature, the huge trade in slaves in the Roman Empire moved large numbers of people great distances. And, as slaves gained their freedom, often settling in places far from where they were born, the ethnic make-up of the empire became very mixed. Tacitus sniffed, '[I]f freedmen were marked off as a separate class, then the scanty number of freeborn would be evident.'

What all of this means is a picture of great complexity and unclarity. If half of all the soldiers who settled in Britain as colonists in places like Colchester were from northern Europe, their DNA will be hard to distinguish in a native British population with pre-existing close links to the opposite shore of the Channel and across the North Sea. And certainly a Mediterranean input in the 400 years of the province of Britannia is impossible to identify with any certainty from current data.

It is possible to count how many lineages in Britain which are common and diverse in Italy and in the Mediterranean generally and use this calculation to give an idea of maximum Roman input. It can be as much as 10 per cent in the south-east of England. However, the same lineages all arise in the Near East and were also carried to Britain by the early farmers fanning out from the Fertile Crescent. Most geneticists consider the latter scenario to be much more important if for no other reason than the fact that many fewer people lived in Britain in the centuries around 3,000 BC and so the effect of the earlier immigrants was concomitantly greater.

If that is true for southern Britain, then it must be emphatically the case in the north as a territory that was only fleetingly and occasionally part of the empire. The maximum figure is 2 per cent but the actual number is likely to have been very much lower. What did the Romans do for Scotland's DNA? Almost certainly very little.

The Iceni uprising appears to have galvanised imperial strategy, such as it was, in Britain. Over the next 20 years, the legions tramped northwards, first subduing the great trans-Pennine kingdom of the Brigantes and then, under the command of Tacitus's father-in-law Agricola, striking deep into Scotland. It seems that the conquest of the whole island of Britain was envisaged. At Mons Graupius in Aberdeenshire, the Romans overcame the Caledonian Confederacy led by the first-named of our ancestors, Calgacus. The men with the red-gold hair and massive limbs described by Tacitus were, like Boudicca's host, no match for the grim, close-quarter fighting of the tight legionary formations and the dash and bravery of their auxiliary cavalry.

With the accession of the Emperor Domitian on a dark tide of rumours about the poisoning of his brother Titus, the army in Britain was ordered to pull back from Scotland. Resources were urgently needed elsewhere and retrenchment became the guiding strategy. In order to secure the northern frontier in Britannia, the provincial government built the Stanegate, the Stone Road, between the eastern fort at Corbridge that guarded the crossing of the Tyne and the western fort at Carlisle that watched the Solway and the ford over the Eden. After the withdrawal from Scotland, more forts were added at half-day marching intervals and, for the moment, Rome seemed content to patrol and be watchful on its most northern horizon.

When Hadrian decided to consolidate further and more perma-nently, building his mighty wall only a mile or so beyond the line of the Stanegate, the military logistics involved were staggering. More than 30,000 soldiers were deployed to source and gather materials, transport them to the site and build. A substantial garrison was posted along the length of the Wall and they were to remain there in various guises and strengths for three centuries. Archaeologists have discovered precious information about precisely who stood on the northern rampart watching for trouble rumbling out of the north. From inscriptions and letters, it is clear that there were units from Tungria and Batavia, themselves frontier peoples from the

Rhine delta and what is now Belgium. Syrian archers were based at the fort at Carvoran and a cohort of Spanish cavalry garrisoned the large base at Maryport, part of the sea wall that ran down the Cumbrian coast – two northern European regiments and two from the Mediterranean stationed on the chilly Solway shore.

When Hadrian's successor, Antoninus Pius, required a fanfare of military glory to usher in his reign in 138, the frontier was briefly moved up to a new wall between the Forth and Clyde, across the narrow waist of Scotland. The Tungrians and Syrians were joined by Thracians, Gauls, the Baetasians from the Netherlands and a cohort recruited in northern Spain. It was an imperial policy to post provincial units some distance away from their origins. Their tour of duty in Scotland was brief and in 157 the line of the frontier was once again pulled back to Hadrian's Wall.

If imperial strategists hoped that consolidation would bring peace, they were quickly disappointed. There was war in northern Britain in 161 and again in 180 when native armies crossed the Wall, massacred a Roman force and killed the governor. The crazy Commodus had succeeded his father Marcus Aurelius and the astute British kings probably hoped to take advantage of weakness and uncertainty in Rome.

Britannia remained vulnerable even though a new governor managed to drive out the invaders. Walls were built around the towns in the south and the Wall garrison strengthened. While Gaul and Spain were fully integrated into the empire, there is a sense that a version of apartheid in Britain lingered. From a population of two to three million, only 10 per cent lived in the hundred or so towns and maybe 50,000 on the villa estates in country districts. When the garrison of the Wall and the legionary fortresses and the inhabitants of the *vici*, the villages attached to them, are taken into account, it may be that only a fifth of the total population of Britannia might be seen as Romano-British. In the countryside the vast majority continued to speak dialects of Old Welsh and to live in a Celtic cultural atmosphere. By contrast, the native languages of Spain and France were dying out and cities were developing. It may be that the central problem for the governors of the province was that, in Britain, Rome simply did not catch on.

After Commodus's inevitable assassination, Septimius Severus established himself. Campaigns in the east, in Parthia, prompted more opportunism from northern kings. The governor of Britannia was forced, in 197, to buy peace with large bribes for the Caledonii and the Maeatae. The historian Dio Cassius had heard of these two powerful peoples:

> The two most important tribes of the Britons [in the north] are the Caledonians and the Maeatae; the names of all the tribes have prac-tically been absorbed in these. The Maeatae dwell close to the wall which divides the island into two parts and the Caledonians [are] next to them. Each of the two inhabit rugged hills with swamps between, possessing neither walled places nor towns, but living by pastoral pursuits and by hunting.

Dio meant the Antonine Wall and one of the swamps was the long, largely unbroken stretch of Flanders Moss, five miles wide and beginning at the foot of Stirling Castle rock and reaching as far west as Aberfoyle. Place-names remember the Maeatae and confirm their location next to the wall. Dumyat rises north of Stirling and it derives from Dun-Maeatae, the fort of the Maeatae. Next to it is Myerton Hill and to the south, near Falkirk, is Myot Hill, perhaps a southern boundary. The kingdom lasted for many centuries. In the seventh century, St Adomnán recounted a famous battle between the men of Argyll led by King Aedan macGabrain and the Miathi. And it seems that their DNA may also have endured.

The rare variant of S145-str43 is found in only 79 bearers in online databases which hold details of up to 200,000 people. Of these, 53 know where their father lines originated – 27 in Scotland, 16 in Ireland and ten in England. The distribution in Scotland is fascinating because it is so localised and it may be that the ghosts of the Maeatae still flit around Stirling. Most men in Scotland with S145-str43 came from Doune, Dunblane, Fintry and Port of Menteith. The dating is right and the clustering more than sug-gestive. The Irish distribution is overwhelmingly in Ulster strongly hinting that bearers arrived during the time of the plantations in the sixteenth and seventeenth centuries and some even share Scottish surnames. There are no markers in the Borders or Lothian but ten in England. A little more history will supply an explanation.

When Septimius Severus at last arrived in Britain in 208 to deal with the threat of the northern kings, he did nothing by half measures. A huge army of 40,000 men – the largest ever seen – marched up Dere Street into Scotland. The Maeatae and Caledonii sued for peace immediately and were ignored. Severus pressed on, scorching the earth, burning and plundering – but resistance did not shrivel. In 211, the kings of the northern kindreds mustered their armies once more. But then fate intervened. At the legionary fortress at York, Severus died and his son Caracalla seized the throne, later killing his brother. Needing to make all haste to Rome to secure a fragile hold on power, he quickly came to terms with the Maeatae and the Caledonii. Usually, this involved the handover of aristocratic hostages to guarantee good behaviour. Such people were often allowed small retinues – and it may just be that these bearers of S145-str43 were accommodated at one of the large forts on Hadrian's Wall. Two of their descendants can trace father lines to Northumberland and another to North Yorkshire. Such links are a stretch, without doubt, but the Scottish data is much more robust. Statistically, this cluster is not insignificant – about 30,000 or 0.6 per cent of all Scottish men carry the marker. It seems that across 2,000 years, the descendants of one of the most warlike lost kingdoms of the north have survived.

The century after Severus's wars appears to have been relatively peaceful despite a rapid turnover of emperors in Rome. Between 235 and 284, 15 reigned briefly and many other pretenders made attempts on the throne with the sporadic and often fickle support of various provincial legions. Against this background of instability, Britain eventually broke away. Carausius, the admiral of the Classis Britannica, the British Fleet, proclaimed himself Emperor of Britain and part of northern Gaul in 286. Seven stormy years later, he was assassinated by the unlikely figure of the civil servant Allectus, his Financial Secretary. During his unsteady reign, the kings in the north began to make plans.

In 296, war bands crossed the Wall and raided far to the south, even attacking the legionary fortress at Chester. An alarmed Roman historian described these insurgents as particularly ferocious and he gave them a new name. Probably first coined as a soldiers' nickname, they were called the Picts – the Painted or Tattooed People. It was a dramatic entrance for one of the great kindreds, the makers of much of Scotland's early history.

While there are no texts and no one could now compile a sentence, whispers of a Pictish language have survived in place-names, particularly those with the prefix *pit* - or *pett* -. In names like Pittodrie in Aberdeen and Pitreavie in Dunfermline, its distribution stretches down the eastern coasts of Scotland to the Forth – pit means 'a portion or piece of land' and is often attached to a personal name in order to identify the owner. For example, Pitcarmick in Perthshire was originally 'Cormack's portion' and Pitkenny in Fife was 'Kenneth's land'. The distribution of *pit* place-names fits closely with the locations of another Pictish legacy. Their wonderful, magical symbol stones are to be found mainly from the coastlands of the Moray Firth down to Fife and sometimes beyond. The names of their provinces or minor kingdoms and lines of Pictish kings have also been preserved but there are very few written records of anything that might be called history. And then, in the ninth century, the Picts appear to disappear.

The beautifully carved but enigmatic symbol stones, the gorgeous jewellery and their sculpture marked the Picts out as a distinct, even exotic culture. But the lack of ethnic evidence of their history has turned an enigma into a mystery. One of the most famous enquiries was entitled *The Problem of the Picts*. Historians have speculated that they were invaders whose culture flourished and then dissolved into nothing, leaving no discernible political or social legacy.

Geneticists can at last offer some clarity and resolve much of the mystery. DNA analysis reveals that the Picts have not disappeared – they are, in fact, still here, living anonymously amongst us. They look like us, live like us and are probably entirely unaware of their extraordinary history, their culture and great art. A marker has been identified that is essentially unique to Scotland and very rarely found anywhere else (there is some small leakage in Ulster, probably as a result of the plantations and a little spreading to the north of England). It is known as R1bstr47 or R1b-Pict and around 10 per cent of Scottish men carry it. In our cities, towns and villages, 250,000 Picts are quietly going about their daily lives.

The distribution of the marker broadly matches Pictish territory and, where later incursions such as the Dalriada Gaels and the Vikings overlaid it, the numbers are diluted. It is well represented in the east of Scotland above the Forth but much less so in the Northern and Western Isles. Skye is an interesting exception. Five early symbol stones have been identified on the island and it may well be that Gaelic arrived there relatively late.

R1b-Pict is at least 3,000 years old and possibly even older and is a subgroup of S145, the marker associated with the Celtic-speaking populations of the Isles. It looks as though the marker developed among the earlier peoples who settled in Scotland and, if recent research is borne out, its bearers are descendants of the first farmers. Therefore, it should be seen as a pre-Pictish marker but one that would have been common amongst the Picts.

The Picts were, of course, related to the British to the south and the Irish to the south-west and their sharing of the S145 marker at high frequencies underpins this. While four centuries of Rome seems to have made little impact on the DNA of Britain, what happened after the collapse of the province after 410 certainly changed our genetic make-up. Slowly at first but with gathering momentum, another group of invaders sailed to Britain and ultimately to Scotland.

Sea raiders from the North Sea coasts of the North German Plain, particularly from the area between the mouths of the Elbe and the Weser and the southern neck of the Danish peninsula, had been attacking the western empire since the third century. Not only had they rasped their boats up on British beaches and plundered settlements for anything portable (including people), they had also raided as far south as the Bay of Biscay. A sophisticated, literary Roman landowner Sidonius Apollinaris was appalled and wrote in the fifth century that these Angles and Saxons were the most brutal, savage barbarians yet seen in the west. When they were about to set sail for home, they were in the habit of offering a sacrifice to their bloodthirsty gods by drowning or crucifying one in ten of their captives thus 'distributing the iniquity of death by the equity of lot'.

The imperial administration in Britannia had been dealing with Anglo-Saxon and other Germanic pirates for most of the fourth century and had built an impressive chain of coastal forts from Brancaster in Norfolk round to Porchester on the Channel shore. For genetic historians the battle lines are less clearly drawn. The garrison of the province had long counted Germanic soldiers amongst its units. At Housesteads Fort on Hadrian's Wall, altars have been found dedicated to gods native to Frisia in north-western Holland. One inscription suggests another Germanic group from even further to the north. The *Numerus Hnaudufredi* translates as 'Notfried's Own' and sounds like a prince and his war band conscripted into the armies of the empire. Much later, in 367, a combined invasion of the province by the Picts, the Scots, a mysterious people probably from the Western Isles known as the Attacotti, the Franks and the Saxons overran Britannia. They attacked the coastal defences, the forts of the Saxon Shore, and killed the imperial forces defending them and their commander. His name was Fullofaudes, a German.

The blurred genetic picture of Germans already in Britannia fighting another set of Germans trying to settle there is matched by an equally unclear sequence of events. It seems that in 406 the army in Britain backed the claims of a usurper-emperor and all that is known of him is his name, Marcus. Very quickly he was replaced by another usurper – this time, a native British candidate called Gratian. And then he, in turn, was removed by Constantine III who was said to have been only a low-ranking soldier.

Meanwhile, the empire was beginning to crumble. After the Rhine froze on the last day of 406, it is thought that more than 70,000 barbarians slithered across the ice and entered Gaul, rampaging, looting, killing and causing chaos. It was a number large enough (and with a datable arrival) to leave a genetic legacy – unlike the small groups of men in small boats who landed on British shores. In 407, Constantine III led what remained of the garrison of Britannia to Gaul to deal with the insurgents. The province was left very vulnerable when a large force of Saxons attacked the eastern coasts a year later. This was a turning point and, in a rare moment of precision, the historian, Zosimus, summed up what happened:

The barbarians [from] across the Rhine attacked everywhere with all their power, and brought the inhabitants of Britain and some of the nations of Gaul to the point of revolting from Roman rule and living on their own, no longer obedient to Roman laws. The Britons took up arms and, braving danger from their own independence, freed their cities from the barbarians threatening them; and all Armorica and the other provinces of Gaul copied the British example and freed themselves in the same way, expelling their Roman governors and establishing their own administration as best they could.

The immediate impact of this break with the imperial past was not clear. Money must gradually have ceased to circulate. Since there had been no mint in Britain for a century, cash to pay soldiers and administrators had to be imported from across the Channel. It reached the length of the province and even beyond Hadrian's Wall. In a field near Kelso in the Scottish Borders, hundreds of Roman coins have been found. The strong suggestion is that they were pay for an outlying unit based in a fort on the mound now occupied by the ruined medieval castle of Roxburgh. Most of the coins are small bronze radiates and some date very late, shading into the fifth century and the reign of the Emperor Honorius, almost to the very last days of the imperial province. The radiates are not in themselves valuable and appear to be evidence of a money economy operating on the banks of the Tweed around 400. Whenever the metal detectorists who found the coins see a new one glinting on the furrows of the ploughed field where most have been turned up, they marvel at the reach of Rome even in its dying days in the west. And when they pick up something dropped by an imperial soldier, they feel a shiver of contact with the long past.

That soldier may have been a German in the pay of the Empire, perhaps hoping for greater rewards as a supporter of a British usurper. Barbarian kings regretted the ultimate fall of Rome, engulfed by the chaos of assaults on the imperial throne by usurpers. Athaulf, the successor of Alaric the Goth who famously sacked Rome in 410, mourned the passing of a lawful society and feared the anarchy that would follow the takeover of the west by barbarian peoples. These kings saw the empire as a career opportunity rather than as a victim to be destroyed but, once the process of disintegration was underway, it became unstoppable.

By the middle of the fifth century, Kent was a Jutic kingdom and Angles and Saxons had begun to settle elsewhere along the coasts of the North Sea. One of their most famous footholds was at Bamburgh, now the site of an immensely impressive restored castle. What may have initially been a base for raiders (the beach allowed boats to be easily dragged up above the high-tide mark and the castle rock offered refuge as well as being a rare seamark along a flat and relatively featureless shore) became the kernel of a great kingdom. By the middle of the sixth century, Bernicia was beginning to emerge.

The name itself is an insight into how this fascinating kingdom came about. Bernicia derives from *Bryneich* or *Berneich* and itis an Old Welsh name that means something like 'the land between the hills'. Given the location of Bamburgh and its closeness to the wide Tweed basin lying between the Cheviot Hills to the south and the Lammermuirs to the north, it looks as though the kings of Bryneich ruled a prime tract of fertile farmland, precisely what was needed to sustain political power in the sixth century.

The rolling fields of the Merse (the name still used for southern Berwickshire and cognate to Mercia which means 'borderlands') were very desirable. Before the age of improvement in the eighteenth and nineteenth centuries, drainage was almost entirely natural, and while flat river valley bottoms are now valued for intensive agriculture, they were often very boggy and unsuitable for raising crops or pasturing beasts. In contrast, the gentle descent of the landscape of the Merse to the banks of the Tweed, with its free-draining fields in undulating and sheltered countryside, was perfect for early farmers. To the eye of a sixth-century landowner, this was a place where rich harvests could be reaped.

The Merse was also part of the Old Welsh-speaking kingdom of Gododdin, with its bases on Edinburgh's castle rock, Traprain Law in East Lothian and probably Eildon Hill North and Yeavering Bell at the eastern edge of the Cheviot ranges. This loose and ill-defined polity pre-existed the collapse of Britannia, having lain to the north of Hadrian's Wall for centuries, and it grew from the kindred known to Roman mapmakers as the Votadini. When it came into violent contact with the Bernicians, it would emerge from the shadows of history for a brief and bloody moment.

Ida, son of Eobba, was the first king to rule from Bamburgh. In his *Ecclesiastical History of the English People*, written in the years around 730, the Venerable Bede noted, 'In the year 547, Ida began his reign, which lasted for twelve years. From him the royal family of the Northumbrians derives its origin.' While clearly a very great historian and a pioneer who checked his sources and actively sought information, Bede was not immune from the pressures of politics. His aim was to compile a history of the English people, the Angles, the Saxons, the Jutes and others who had overrun England by the time he sat down to write. He was not much interested in the natives, the vast majority of the population. Instead, royal ancestry, legitimacy and continuity were a central part of his purpose and Northumbrian authority and its hegemony over the rest of England was portrayed as a pleasing and civilised continuation of Roman rule.

There are few alternative written sources for the Dark Ages in Britain and one of the most fascinating and maddening is Nennius's *History of the Britons*. Amongst much myth-history and tales of giants and dragons, there is a series of related passages known as the North British section. Scholars have argued that these are fragments from a lost chronicle of the northern kingdoms of the fifth and sixth centuries. It may have been the work of monks at Whithorn in Galloway or Glasgow. In any event, it states that 'Ida joined Din Guauroy to Berneich'. Din Guauroy was the original Old Welsh name for the fort at Bamburgh (renamed after Ida's queen, Bebba) and Berneich the earlier version of Bernicia. There is none of the usual language of conquest or war in either Nennius or Bede and it may be that Ida did indeed effect the union of two kingdoms. Perhaps he led a small but vigorous war band who established their leadership of the men of neighbouring Berneich, a Celtic kingdom, perhaps a southern or satellite province of Gododdin. From this union flowered Northumbria, the first great power of the Dark Ages in northern Britain.

Despite its political prowess and cultural achievements, there is very little genetic evidence for significant Anglian immigration in the north or in southern Scotland. But that is not to say that there is no signal at all. The S21 group is more frequent among south-eastern Scottish Y chromosomes and, in fact, a subgroup labelled S29 is found there – and it has not been detected anywhere else in Scotland. It is very much an English marker in the British Isles, concentrated in the east of England and also spread widely on the North European Plain. Tantalising hints like this do suggest that these are traces of Anglian DNA. Most other Germanic genes from the period are, at present, difficult to distinguish from earlier arrivals and more definite estimates will have to wait for more markers to be identified in the future. However, the signal in the Lothians and the Borders is much lower than that in East Anglia. Quite probably, therefore, a huge influence was exerted by a small group of powerful incomers over a large native population. When the British ruled the subcontinent of India and Pakistan, they did so with a tiny garrison and, although they were better equipped, the natives could have grouped together and overcome them. But they did not.

The most likely solution to the mystery of Anglian domination is fusion. Encouraged by Bede, the Arthurian stories and modern historians reading history backwards, we too easily see a Celtic–Anglian divide, a struggle between the original British and ambitious bands of incomers. Such evidence as exists suggests a more complex picture with natives fighting in Anglian armies and vice versa. What interested the kings of the sixth century was power not ethnicity.

With the departure of the legions and the remnants of the imperial administration of Britannia, the apparatus of government quickly faded away. Bureaucracy, record keeping, tax collection, a money economy, organised trade, the upkeep of services and the maintenance of towns by their agricultural hinterland began to crumble and fragment. Power shifted towards kings, mostly native aristocrats who had never lost their identity or prestige throughout the long centuries of Roman occupation and who commanded war bands to enforce their authority. It was personal and often fleeting, rising and falling with the death or success of individuals in warfare. Kingdoms were not formal administrative entities with frontiers and differing cultures. Instead, they formed around charismatic and aggressive kings and were measured only by the extent of their military reach.

For these reasons, a political geography of Scotland and North Britain after 410 can only be provisional, subject to continual modification and, in any case, based on little more than scraps of historical evidence. Or it can be entirely mysterious. One scrap talked of an early Anglian presence on the fringes of Berneich in the second half of the fifth century. Nennius noted that a war band led by a man called Eosa had established itself somewhere between the Tyne and the Tweed and, more, that he was the progenitor of Anglo-Saxons who would be kings. Another tradition asserts Eosa as the grandfather of Ida of Bamburgh.

These warriors may well have been descendants of *foederati*, mercenaries imported by the imperial government in the fourth century to protect Britannia from the raids of the Picts and the Scots from the north. More material from the *History of the Britons* adds to this impression. Northern *foederati* based in the Vale of York and led by a man called Soemil were said to have been Angles. Ambitious and vigorous, they 'first separated Deira from Bernicia'. Corresponding roughly to the old county of Yorkshire, Deira had been a native kingdom with a name something like Deor, Dewr or Deifr. It gave its name to Dere Street, the Roman road that began at the gates of York and led north as far as Edinburgh. By 'separated', Nennius probably meant a takeover, with Soemil and his men detaching the kingdom of Deira from a greater Bernicia and assuming control. This probably happened in the early sixth century and the accession of Ida at Bamburgh in 547 appears to follow a pattern repeating all over the old province.

If these early northern kingdoms were indeed the joint creations of native and Anglian aristocrats and their war bands, they were always led by men who bore English names. That does not necessarily mean that they were, in fact, ethnic Angles, only that Anglian culture and language were in the ascendant. Military prowess may have been the blunt impetus behind that and, while records of who fought for whom and for what are very scant, one example intrigues. When the armies of the Gododdin, the kingdom of the

Lothians and probably the Tweed Basin combined with their allies and rode south to meet an Anglian host at Catterick around the year 600, they were led by Yrfai map Golistan, Lord of Edinburgh. His name looks and sounds Celtic but, in fact, the second element is only a thin disguise and it means 'son of Wulfstan'. The sources say that Yrfai was not a nobleman and it seems that, instead, he was a soldier of Anglian descent, perhaps a professional commander.

The takeover of Britannia and its patchwork of Celtic kingdoms in the two centuries after 450 is a remarkable story given the relative numbers involved. A native population of two to three million was subdued by a series of incoming war bands of no more than 200,000 in total who arrived over a long period. How they achieved so much so quickly might be explained in part by the genetic evidence.

As a result of a process coyly termed social selection, scientists have identified an old lineage in Ireland dating from around 400 to 500. Known as M222, it is astonishingly common. No less than 20 per cent of all Irish men carry it! Its distribution is heavily weighted to the north with 40 per cent in Ulster, 30 per cent in Connaught and 10–15 per cent in Munster and Leinster. No less than a fifth of all Irish men are directly descended from one man who lived around 1,500 years ago.

Given the distribution of the marker and its bias to Ulster and especially to men with the O'Neill and O'Donnell surnames, there exists a clear candidate. The Ui Neill kindred dominated Irish history from the fifth to the tenth centuries and their founder was the High King known as Niall Noigiallach. His political reach is reflected in his second name for Noigiallach means 'of the Nine Hostages'. These were the sons of lesser kings who owed Niall obedience and they were kept in his retinue as a guarantee of continued obedience.

The simple reality is that Niall fathered many sons on many women and those sons, themselves growing up to be powerful men, followed the same pattern. Niall's reign and his exploits are shrouded in mist and mystery but, in the historic period, one of his ancestors showed how it was done, so to speak. Lord Turlough O'Donnell, who died in 1423, carried on the family tradition with gusto. He had 14 sons and 59 male grandchildren.

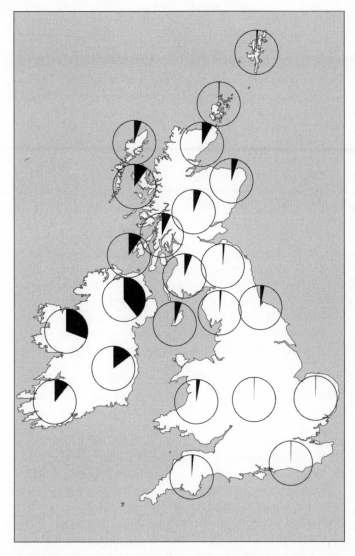

The frequencies of the M222 Y chromosome group are shown across the British Isles using pie charts. Up to 3,000 samples were used to create this map.

From this example alone, it is easy to see how a marker multiplied very quickly. If the same level of enthusiasm and fertility were sustained, Lord Turlough would have had 248 great-grandsons and 1,040 great-great-grandsons. Within only four generations and without taking any account of bastards, he could have bred an army.

The most spectacular example of social selection was discovered in central Asia. When a survey of the DNA of more than 2,000 men was carried out, the researchers found a large group of markers which were very closely related. More than 8 per cent of the entire sample, across 16 populations in a huge geographical area from central Asia to the Pacific coasts, carried the same marker. It was possible to date its origin to about 1,000 years ago. The second-highest frequency of the marker and the place where its internal diversity was greatest turned out to be Mongolia, undoubtedly the place it came from. The lineage is unique and it is unquestionably the case that 8 per cent of all men from Uzbekistan to Manchuria, a total of 16 million, 0.5 per cent of all men on Earth, are descended from one man. And, outside this area, there are no men carrying this marker. Something remarkable has happened in the past.

Over only 34 generations, this is beyond anything seen in nature amongst animals as a result of natural selection. Statistical calculations rule out chance as an explanation. It must be the case that, not only has a dramatically sustained streak of the reproductive fitness of men with this marker been observed, there must have been large-scale elimination of other Y chromosome lineages going on at the same time. There can be only one candidate for the progenitor of 16 million men in central Asia, 0.5 per cent of all men on the planet, and that is the great warlord, Genghis Khan. He lived between c.1162 and 1227 and he and his brothers created the largest land empire in history, often slaughtering the conquered populations as his hordes rode over the plains and attacked cities and other tribes. The boundary of his empire at its fullest extent is almost exactly matched by the frequency of his marker. There is only one exception – the Hazaras of Pakistan. Their traditions claim Genghis as their ancestor and, in fact, the highest frequency of the marker is found amongst the Hazaras. It exists nowhere else in Pakistan.

Much closer to Scotland was a man who died when the great Khan was born. His exertions bore a good deal less fruit but then he operated in a much smaller area. More than 50,000 Scottish men, most of them with the surname of MacDonald or its variants, are the direct descendants of Somerled. The first Lord of the Isles and founder of Clan Donald, he ruled the Hebrides and was King of the Isle of Man. When he clashed with Malcolm IV of Scotland at Renfrew in 1164, Somerled was killed – but his genes certainly lived on.

The alleged medieval tradition of droit de seigneur – or the right of a local lord to have sex with any new bride in the community he controlled – was also known as the *ius primae noctis* or the 'right of the first night'. The idea that a lord might be powerful enough to insist that he should deflower a new bride before a new husband has long been thought little more than a male fantasy. But DNA studies do suggest that something akin to most teenage boys' wildest dreams did go on throughout history. While it is unlikely that seigneurs exercised their droit in quite that way, with each new bride, it is certain that they had productive sex with many women. Perhaps there was an informal *ius secundae noctis*.

When the church began to promote ideas of legitimacy and illegitimacy, it may have been to use these to act as a brake on what it saw as excessive lordly licence with the women in any community. Perhaps priests and monks saw such practices as potentially very socially divisive. Nevertheless, in the historic period, the chiefs of the highland clans made little meaningful distinction between the offspring of their wives and so-called bastard children. Somerled must have adopted a similar attitude. And early Irish law tracts encouraged the fathers of children 'begotten in brake or bush' to recognise them as their own.

If social selection went on on a spectacular scale with Genghis Khan, Niall Noigiallach and more modestly with Somerled, then it must certainly have been the norm locally. Some researchers who identified the marker R1b-Pict believe that the cluster of similarities is like that of the Genghis marker, although not so tight or so recent. It is probably a signal of social selection in the deeper past and perhaps those who carry it – around 10 per cent of Scottish men – are the descendants of prehistoric royalty who ruled north of the Forth and who were the progenitors of the Pictish nation.

Ida of Bamburgh was said to have had six sons and, in an age when the effects of a lack of contraception were balanced by the number of mothers and babies who died from complications in childbirth, that is not an unreasonable number – restrained, in fact, compared with the efforts of Lord Turlough O'Donnell. In any event, the multiplying arithmetic of power is not difficult to visualise. It may well be that the Angles, Saxons, Jutes and other Germanic groups who sailed to Britain in small boats simply out-bred as well as out-fought the natives. And, if there were massacres on any scale (some are reported but the characteristic hyperbole of chroniclers is likely to have inflated the numbers of dead on the battlefields and elsewhere – Aethelfrith of Bernicia was said to have slaughtered 1,200 monks at Bangor), then the effects of social selection might have been as sweeping as they were in central Asia, even if the numbers involved were very much smaller.

With the defeat of the Gododdin armies led by Yrfai map Golistan at Catterick in 600, the next stage for the drama of Scotland's developing story was set. Led by Aethelfrith and the descendants of Ida, the Bernicians began to push aggressively northwards and they would eventually clash with the Picts. The ancient hegemony of the Old Welsh-speaking kingdoms of the north was shrinking and would quickly coalesce around Strathclyde, a remarkable survival story. And, in the west, a new power was gathering strength. The war bands of the kings of the Scots would ride east and their ambition and acuity would ultimately confer their name on the whole country.

8

The Four Nations of Scotland

✖

HERE IS PART OF THE opening chapter of Bede of Jarrow's *Ecclesiastical History of the English People*:

At the present time there are in Britain, in harmony with the five books of the divine law, five languages and four nations – English, British, Irish and Picts. Each of these have their own language; but all are united in their study of God's truth by the fifth – Latin – which has become a common medium through the study of the scriptures. At first the only inhabitants of the island were the Britons, from whom it takes its name, and who, according to tradition, crossed into Britain from Armorica, and occupied the southern parts. When they had spread northwards and possessed the greater part of the island, it is said that some Picts from Scythia put to sea in a few longships, and were driven by storms around the coasts of Britain, arriving at length on the north coast of Ireland. Here they found the nation of the Irish, from whom they asked permission to settle; but their request was refused . . . The Irish replied that there was not room for them both, but said: 'We can give you good advice. We know that there is another island not far to the east, which we often see in the distance on clear days. If you choose to go there, you can make it fit to live in; should you meet resistance, we will come to your aid.' So the Picts crossed into

Britain, and began to settle in the north of the island, since the Britons were in possession of the south. Having no women with them, these Picts asked wives of the Irish, who consented on condition that, when any dispute arose, they should choose a king from the female royal line rather than the male. This custom continues among the Picts to this day. As time went on, Britain received a third nation, that of the Irish; they migrated from Ireland under their chieftain Reuda and by a combination of force and treaty, obtained from the Picts the settlements they still hold. From the name of this chieftain, they are still known as the Dalreudans, for in their tongue 'dal' means a division.

An uncharacteristic mix of history, myth-history and speculation, Bede's summary of the origins of the British is fascinating. Although he could not have known it, the tradition that the first settlers, the pioneers, came from the south is, of course, borne out by DNA studies. The subsequent arrival of the Picts from Scythia then veers wildly into the shadowlands of giants, dragons and ancient epics. By Scythia, Bede probably meant northern Germany. Other historians made the same link and it may have denoted no more than a wide area of barbarian Europe – that part of the continent outside of the Roman Empire. As the author of the *Ecclesiastical History of the English People*, it was attractive to Bede to connect the fabled, gilded and warlike Scythians with the Angles, Saxons and Jutes and, more, to set down a precedent for their incursions into Britain. In historiography, the past could often be used to legitimise the present.

The charming account of a conversation between the migrating Picts and the Irish somewhere on the Ulster coast gave rise to persistent mythology. The descent of Pictish kings through the mother line added an exotic tinge to the deepening mystery of the origins of the early peoples of northern Scotland beyond the Forth and Clyde. It was a tale much repeated over many centuries. And the Picts were painted as incomers, a breed apart, strange and different from the original inhabitants of Britain. It turns out that the genetic and historical realities were very different.

Where Bede's famous account shades back into history is in

the section on the beginnings of Dalriada (or Dalreuda) and the Gaelic-speaking kings of the Atlantic west. Recent scholarship has made a sensible assertion that the people living on both shores of the North Channel were probably cousins, related communities who shared the Gaelic language and possibly formed part of a coherent sea kingdom. Until the coming of the railways and the roads in the nineteenth and twentieth centuries, the sea was properly seen as a connector rather than a divider.

Walter Bower, one of the first students at St Andrews University soon after it was founded, compiled the *Scotichronicon*, a history of Scotland. Writing in the early fifteenth century, he drew on the work of Bede of Jarrow and incorporated much material collected by the scholar, John of Fordun, 50 years earlier. The opening chapters of the *Scotichronicon* deal with the tale of Scota, daughter of the Egyptian Pharaoh, Chencres. She married Gaythelos who was good looking but mentally unstable. Despite the latter and probably because of the former, Scota became devoted and, when a slave revolt drove the happy couple out of Egypt, they fetched up on the banks of the River Ebro in the Iberian Peninsula. Quoting from the legend of St Brendan, which leaned heavily on the *An Lebor Gabala Erenn*, the handsome but daft Gaythelos was said to have established a kingdom. There, his descendants multiplied greatly and eventually set sail for Ireland. The geography appears confused but the direction of travel familiar. In a dubious legacy, the mentally unstable Gaythelos allegedly gave his name to the Gaitheli or the Gaels and, of course, Scota gave hers to the Scots.

Bower then takes his readers on a tour of the islands around Scotland's coast, ending with his own. He was the Abbot of the Augustinian Canons and the beautiful ruins of their church still stand on the island of Inchcolm in the Firth of Forth. Following Bede's lead on the origins of the Picts out of Scythia (but this time correctly locating it on the shores of the Black Sea so that he could neatly make an early connection between the Scots and St Andrew) and including the material of matrilineal succession, Bower then offered something more concrete and a great deal more likely. After the Picts had crossed the North Channel with their Irish wives in tow and begun to settle in Argyll and probably elsewhere:

[t]hey were followed by countless numbers of their kinsfolk, fathers and mothers, brothers and sisters, nieces and nephews, and also many others who were not only motivated by love for a daughter or sister, but rather they were very strongly attracted by the grassy fertility and abundant pasture for their herds in the land of Albion for which they were heading. The number of people of both sexes following them and taking their cattle with them who went off in a short space of time to live with the Picts was larger than has ever been recorded as having left their own land without a leader. But their number was also increased by an endless succession of criminals, because anyone who was in fear of incurring the penalty of the law went off to live with the Picts scot-free. Then they sent for their wives and children and remained there peacefully never again to return.

The key phrase in that interesting passage is 'without a leader'. Was this a fuzzy but embroidered account of a folk migration? Certainly Irish war bands had raided western Britain in the last centuries of the Roman province and after 400 had begun to settle. Inscriptions in the Irish Ogham alphabet (like runes but incised vertically on the edges of standing stones) track the incursions of colonists, for these are usually names and are interpreted as boundary markers. They have been found in numbers in Cornwall, West Wales, North Wales, the Isle of Man, Galloway and an isolated example as far east as Selkirk in the Scottish Borders. They also appear in Argyll, Lorne and Cowal, the part of Atlantic Scotland that became Dalriada.

At first, a sea kingdom linked by the North Channel and incorporating Antrim as well as Kintyre, Arran and Argyll, it developed out of the territory of the great Irish dynasts, the Ui Neill. When Walter Bower's narrative arrived at the coming of the kings of Dalriada to Scotland, he added a little helpful background. Fergus, son of Feraghad or Ferard, heard that 'a tribe of his own nation without anyone to lead or govern them was spending their time wandering through the desert wilderness of Albion . . . he blazed with anger in his heart!' Also attracted by reports that Scotland/

Albion was indeed fertile, Fergus crossed the North Channel with a war band and made himself king.

Perhaps a descendant, certainly a historical figure, Fergus Mor mac Erc ruled over Dalriada some time around 500. His particular significance may have been in a shift of political focus for Fergus appears to have been based in Argyll and not Antrim. Early sources talked of three sons of Erc conquering in Scotland. Loarn took the area of northern Argyll now named for him as Lorne while Oengus or Angus colonised Islay and Jura, and the principal fortress of Fergus was almost certainly the spectacular Dunadd, a singular rock rising out of the great marshland around it. As a dynasty deriving from Ireland, it is likely that their political structures reflected those of their homeland.

The Gaelic word for king is *righ* or *ri* and, in Ireland, there existed three sorts. The lowest rank was local – a *ri* or *ri tuaithe* who ruled over a kindred or small kingdom. He owed submission to a *ruairi*, an overking, who was based in his own kindred but had grown more powerful. This kind of relationship was probably dynamic if it depended on force or arms as well as tradition. The *ri ruirech*, the High King, ruled over all kings. The energetically fecund Niall Noigiallach was one of the first High Kings of Ireland and the second part of his name makes reference to the nine kings who submitted. The powerful Ui Neill dynasty produced long lines of High Kings and the title endured for many centuries. Perhaps its most famous bearer was Brian Boru, who styled himself 'Emperor of the Irish' in 1002 and defeated the Vikings at the Battle of Clontarf in 1014.

Fergus Mor mac Erc was no emperor but almost certainly overking of Dalriada. Loarn and Oengus may indeed have been his brothers but they were each a *ri tuaithe* in Islay, Jura and Lorne. Fergus himself held Kintyre. A good deal is known about the kindreds of Dalriada because of a fascinating document known as the *Senchus Fer nAlban*, the *History of the Men of Scotland*, meaning not the whole country but only that part settled by the Scots.

In essence a muster roll for a naval power, it lists each kindred and the number of 'houses' or farmsteads it contained. The purpose of the *Senchus* was to reckon the number of ships that could be launched in wartime and also the force of marines needed to row them. Here is the table:

Cenel nOengusa (the kindred of Angus on Islay and Jura) 430 houses
Cenel Loairne (the kindred of Lorne in Lorne and Appin) 420 houses
Cenel nGabrain (the kindred of Gabrain in Argyll) 560 houses

There was a good deal of regional variation, some of it recognised by the compilers of the *Senchus* ('small are the lands of the houses of the Cenel nOengusa') but it looks as though a group of 20 houses was bound to provide 28 oarsmen, enough to row two of the larger sea-going curraghs, which were called sevenbenchers. When all of his forces were mustered, the Overking of Dalriada could command a powerful fleet – more than 70 curraghs with almost 1000 marines on board. What is striking is the degree and precision of military organisation, surely a sign of driving royal ambition. Walter Bower noted that Fergus Mor mac Erc had indeed expanded his kingdom, ruling beyond Drumalban – that is, beyond the ridge of Albany. He meant the long and forbidding chain of high mountains that rose up at the head of Loch Lomond and stretched north to Ben Nevis and the Great Glen, what Adomnán called 'the Mountains of the Spine of Britain'.

Were Fergus Mor and his Dalriadic kingdom different? A new element in Scotland's genetic make-up? The S145 marker, which scientists have characterised as the quintessential marker of Celtic language speakers, may have its origins around 1,000 BC or before and it was shared by the peoples on both sides of the Irish Sea and the North Channel. But, as the molecular clock ticked on, mutations appeared, amongst them the spectacular legacy of Niall Noigiallach, the M222 marker. Did it cross the sea with the war bands of Fergus Mor mac Erc and his ancestors?

There is uncompromising evidence that it did – and in very substantial numbers. More than 6 per cent of all Scottish men carry M222, around 150,000 are direct descendants of Niall, the High King of the Irish. The frequency of the marker is very pronounced in the west with 9 per cent and less in the east with

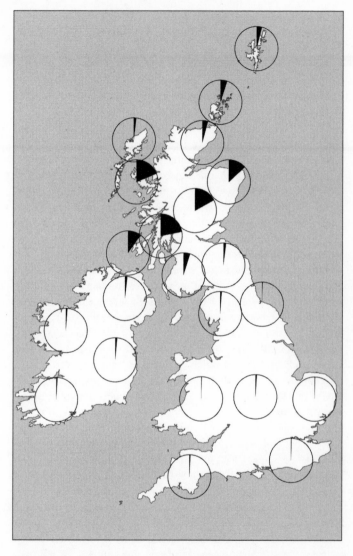

The frequencies of the S145-str47 'Pictish' Y chromosome group are shown across the British Isles using pie charts. Up to 3,000 samples were used to create this map.

3 per cent on an axis from Galloway to Shetland. It occurs very often amongst men with ancient Scottish surnames whose family trees can, in some cases, be traced back over three centuries. Those in Scotland with the M222 marker are not recent immigrants and their high incidence and geographic spread indicate a large-scale movement of people – probably mainly from Ulster and probably around AD 500.

The clear descent from Niall Noigiallach implies that around the time of Fergus Mor mac Erc it had not yet dissipated and was carried by men of high status. These were almost certainly warriors, the leading men in war bands, and they had territorial ambitions. Perhaps they were groups of younger sons pushed out to make their own way as elder brothers claimed inheritance. It is a pattern that is clearly observable at several points in European history, particularly in the Norman invasion of England in 1066 and the time of the Crusades.

Other Irish-specific markers from the period around AD 500 can be found in Scotland and their presence reinforces a sense of colonisation. S168 is relatively rare and strongly concentrated around the mouth of the River Shannon where it is now found in Tipperary and Limerick. This was once the territory of the Dalcassian clans, the descendants of the great High King Brian Boru. But it does occur rarely in Scotland and as far east as Montrose. S169 is most common in Leinster, the lands of the Lagin clans, and it too is found in Scotland, especially amongst men with the surnames of Beattie and Ferguson.

The genetic and political divide between incomers and natives also had cultural facets. The Dalriadans spoke Irish Gaelic, Q-Celtic, while the Picts spoke P-Celtic and it appears that the languages were not mutually intelligible. When St Columba attempted to bring the Word of God to the Picts, it had to be translated. Scholars believe that Pictish was more closely related to the speech of mainland Britain, what might be called Old Welsh. That meant a greater affinity with the people Bede identified as the British even though he clearly saw Pictish as a separate language. Perhaps, by the early eighth century, Old Welsh and Pictish had lost their similarities and grown apart.

There is also a sense of Pictland harking back to a British past, a culture that preserved some of the early characteristics of Britishness. Roman commentators remained interested in the Picts for several centuries perhaps because of their reputation for atavistic ferocity. Here is Herodian writing in 235 with a fascinating observation:

> They are ignorant of the use of clothes . . . they tattoo their bodies not only with the likenesses of animals of all kinds, but with all sorts of drawings. And this is the reason why they do not wear clothes, to avoid hiding the drawings on their bodies.

There may well have been a religious reason why they did not hide their markings – the gods needed to see them, as did their enemies. The name of Britain derived from a Greek coining, Pretannike, which meant The Land of the Tattooed People and the literal meaning of the label was persistent. In the late third century, Tertullian wrote of the 'stigmata Britonum' and, later, the poet Claudian saw a woman with her cheeks tattooed as a symbol for Britain. He confirmed that this was more particularly a Pictish practice when he praised the Roman army as 'this legion which curbs the savage Scot and studies the iron-wrought designs on the face of the dying Pict'. But the habit of tattooing was waning in the south and the original meaning of the name of Britain was gradually forgotten. Only in the far north, beyond the old limits of the Empire, did ancient cultural habits die hard. By 600, Isidore of Seville was implying that tattooing was now important to only one people:

> The race of Picts have a name derived from their bodies. These are played upon by a needle working with small pricks and by the squeezed-out sap of a native plant, so that they bear the resultant marks according to the personal rank of the individual, their painted limbs being marked to show their high birth.

Surprisingly, some of these marks or designs appear to have survived. The Picts left little or no literary evidence of their culture and history and few other material remains have come to light. But one dazzling phenomenon does stand testament in the landscape – their great symbol stones remember the Pictish centuries in the north of Scotland and they tell a rich story.

Two hundred stones have survived, almost certainly a small fraction of the original number. They may have been used as boundary markers, memorials or simply as expressions of personal prestige. The earliest group are known as Class I and, pre-Christian, they date to the time before 600. Many carry mysterious abstract symbols as well as representations of animals. Following Isidore's observations, these may well have been tattooed on the bodies of individuals and were full of significant, possibly religious, meaning. Many cultures have used animals as totems and early Roman maps listed peoples in northern Scotland with suggestive names. The Epidii in Kintyre translate as the 'Horse Kindred', the Lugi in Sutherland were the 'Raven People' and the Orcades in Orkney, the 'Wild Boar People'. Birds, horses, boars and other animals appear on the symbol stones and, at Burghead on the southern shore of the Moray Firth, the old Pictish naval fortress seemed to be dedicated to a bull cult. Thirty carvings of bulls were found on the site although only six now survive. There were other echoes around the firth. Roman mapmakers plotted a coastal fortification called Tarvedunum, the Bull Fort, and Thurso translates as Bull's Water.

By comparison, the abstract symbols are not easy to read but they may relate to a prehistoric religious rite where metalwork and especially broken weapons were deposited in watery places. Many have been found in lochs and bogs all over Scotland. Carvings on the stones, known as Z-rods, V-rods and the so-called 'tuning forks', may well be stylised representations of broken spears, broken arrows and broken swords. For some reason now lost, weapons were slighted or damaged prior to being thrown in water or buried in marshy ground. Other symbols, such as a crescent and a double disc, could be seen as a torc and a burial chariot. As a development of the practice of offering these objects in sacrifice, Pictish society may have taken to erecting stones carrying symbols of them so their religious devotion might become more conspicuous in the landscape. Like Christian crosses, these pagan monuments were an advertisement of belief in the old gods and what they demanded of mortal men.

The distribution of the stones, as with the *pit-* names over the north of Scotland, suggest a Pictish society spread wide, from the Northern Isles down to Fife and west to the Hebrides but probably excluding Gaelic-speaking Argyll. Later but credible traditions have left a list of so-called Pictish provinces. These were more likely small or sub-kingdoms and here is a translation from the original Irish by the great toponymy scholar, W. J. Watson:

Seven of Cruithne's children divided Alba into seven divisions,
The portion of Cat, of Ce, of Cirech, children with hundreds of
    possessions,
The portion of Fib, of Fidaid, of Fotla, and of Fortriu.
And it is the name of each man of them that is on his land.

Using a twelfth-century recension, Watson went on to tidy up the list and link it with modern counties and districts:

Cirech (also written as *Circenn*)	Angus and the Mearns
Fotla	Atholl and Gowrie
Fortriu	Strathearn and Menteith
Fib	Fife with Fothreve
Ce	Marr and Buchan
Fidaid	Moray and Easter Ross
Cat	Caithness and south-east Sutherland

With the likely translation of Fothreve as Kinross, the pattern appears to reflect a recognisable historic arrangement of a *ri tuaithe* and a *ruiri* – a local, lesser king and an overking. These seven territories seem to divide into two groups with the top four being distinct from the bottom three.

One of Scotland's greatest – and largely forgotten – geographical barriers was The Mounth, where the Grampian massif almost reaches the sea at Stonehaven. In the tenth century, this area was known as the *Claideom* or the 'Swordland'. The name was more common in Ireland and was used for an area of dispute, a boundary land between kingdoms. It may well be that Pictland divided into a northern high kingdom based around the Moray Firth and a southern with its focus on the ancient land of Fortriu in the beautiful and fertile valley of Strathearn.

By the twelfth century, the English chronicler Henry of Huntingdon noted that of all the peoples of Britain listed by Bede only the Picts had perished. It was the beginning of the great mystery. Where had they gone? Nowhere, according to geneticists, and, in fact, the symbol stones and carvings of the later period, the Class II of the Christian era, are eloquent. They show an aristocracy interested in hunting, in finery and horses – a vigorous culture. Found on the Brough of Birsay in Orkney, there is a carving of what appears to be a royal procession. The leading man of three may well be the Pictish king of Orkney, the earliest portrait of a ruler in Scotland – and by some distance. Not only does their DNA marker of R1b-Pict survive widely, the stones offer some concrete sense of what these people looked like. Nothing similar exists for any of the other peoples mentioned by Bede.

In the churchyard at Aberlemno in Tayside stands one of Scotland's most impressive historical documents. On one side of the famous Pictish stone is carved a sequence of scenes from one of the most pivotal battles ever fought in the north. The effects of its outcome still ripple down the centuries but, first, its context needs to be quickly sketched.

As Walter Bower wrote, the power of the Argyll kings had reached over the Drumalban Mountains and into Pictland. The names of two of the Pictish provinces or kingdoms listed by W. J. Watson appear to remember settlers from the Gaelic west. Atholl derives from Ath-Fotla which means 'New Ireland' and its pendant territory Gowrie is from Cenel nGabrain, the name of the kindred of Kintyre that supplied most of the Dalriadan kings.

In 574, the beginnings of more eastwardly expansion may have begun to germinate when St Columba was involved in the installation of Aedan macGabrain as the overking of Dalriada. Perhaps there was a hope that Christianity would spread eastwards as the Gaelic-speaking war bands encountered the heathen Picts. Aedan was very ambitious, campaigning deep into Pictland and as far north as Orkney. But his momentum came to a juddering halt in 603 at a place called Degsastan. After raiding into the Tweed valley, the newly acquired territory of the Bernicians, Aedan's war band made their way back north through Lauderdale. Aethelfrith and his men caught up at the great hill fort above Addinston Farm and a bloody fight ensued. The king of the Bernicians prevailed and, although he appears to have escaped with his life, little more was heard from Aedan mac-Gabrain. It was to be a lengthy but temporary setback for the ambitions of the Dalriadan kings.

By the 630s, the Bernicians had stormed the citadel of the Gododdin on Edinburgh's castle rock and, in East Lothian, a rash of new English place-names began to establish themselves. Whittingehame was the settlement of the people of Hwita, Haddington, that of the people of Hoedda and Tyninghame, the settlement by the River Tyne. The -*ham* names are also Anglian and new arrivals came to Auldhame, Morham and Oldhamstocks. Some time after 640 the Bernicians took the place Bede called Urbs Giudi, the fortress on Stirling Castle rock and they began to edge further north into Fife, the Pictish kingdom of Fib and Fothreve.

Kings of Northumbria, a potent combination of Bernicia and Deira (much of modern Yorkshire) grew strong in the north. Able to manipulate the royal succession in southern Pictland to ensure a series of puppet rulers, they seemed to be unstoppable. But the death of King Oswy saw Northumbrian influence slacken for a short time and their placeman, King Drest, was removed. The Pictish nobility supported Bridei mapBili even though he was a son of the king of Strathclyde. It seems that he had an ancestral claim through a Pictish grandfather, Neithon. In any event, the new king of the Northumbrians, Ecgfrith, did not hesitate. Moving north with a powerful force, he confronted the Pictish host somewhere between the rivers Carron and Avon in 672. He destroyed the army of Bridei but did not kill or capture him and, in the decade afterwards, the Pictish king rallied support and was defiant. When Ecgfrith was forced to lead his cavalry north once more to assert his authority in 685, the stage was set for the stirring scenes on the Aberlemno Stone.

Depending on perspective, that battle fought near Forfar in 685 could have different names. The Anglians called it Nechtansmere, 'Nechtan's Lake' while the Picts knew the place as something like *Dun Nechtain* or 'Nechtan's Fort'. To speakers of Old Welsh like King Bridei mapBili, the place where the armies clashed was *Linn Garan,* 'the Heron's Lake'. Perhaps because of its outcome, historians have tended to follow the Pictish form and they usually record it as Dunnichen. It had a profound impact on all three speech communities.

The memorial at Aberlemno shows what happened in four scenes arranged apparently in chronological sequence from top to bottom so that they can be read like a newspaper comic strip. In his description of events, Bede wrote that Ecgfrith, the Northumbrian king, was lured into narrow mountain passes by the Picts who pretended to retreat. A familiar tactic, it led to a trap and an ambush being sprung and, if the opening scene on the stone is read correctly, it shows a Pictish cavalry warrior chasing a Northumbrian horseman who had jettisoned his sword and shield. Perhaps it is Ecgfrith fleeing in panic from Bridei.

Below this, a second scene seems to show the battle at its height, with a Northumbrian cavalryman charging a tight formation of Pictish infantrymen (their different styles of helmets distinguish the two sides). They appear to be drawn in three ranks like Roman legionaries and the sculptor understood exactly how and where each soldier placed their weapons. The man in the front rank has his shield raised and his sword resting on his shoulder ready to strike while the man behind him holds a long spear with both hands. He has pushed it out beyond his comrade so that it presents a bristling deterrent to the enemy and also defends the front rank. In the rear, a spearman stands with his spear planted on the ground, waiting in reserve.

The third scene looks like a climactic last act as two cavalrymen face each other. Again it is tempting to see them as the two contending kings. The Northumbrian is about to throw his spear while the Pict readies himself to deflect it, his shield held well up to protect his head and shoulders. And, in a final scene, the aftermath is tucked into the bottom right hand corner of the stone where a

Northumbrian warrior lies dead and a raven pecks at his eyes.

Chroniclers record a devastating defeat for the invaders at Dunnichen and that Ecgfrith was killed on that day, 20 May 685. Victory allowed Bridei to re-establish the independent realm of the southern Picts. Bede was not sympathetic to Ecgfrith. Because he had ignored the advice of St Cuthbert, the Anglian king had got little more than his just deserts. Aldfrith, his successor, ably recovered the destroyed state of the realm, albeit within narrower limits, said the *Ecclesiastical History of the English People*. After Dunnichen, the northern frontier of Northumbria probably ran from the Ochil Hills in a diagonal down to the shores of the Solway Firth in the south-west. Encompassing much of the most fertile land in Scotland – at least to the eye of a seventh-century farmer – it was a sufficiently large and influential realm to establish the English language in Scotland, lasting long enough for the incoming language to put down permanent roots.

In the seventh century Dumfriesshire was absorbed into Northumbria despite the disaster at Dunnichen, and archaeologists have found traces of a large monastery at Hoddom on the River Annan. It was probably re-founded by Oswy, Ecgfrith's predecessor. It may have been endowed in memory of Rieinmelth, Oswy's first wife. She was the great-granddaughter of Urien, the king of Rheged.

In the two centuries after the fall of Britannia, several powerful Old Welsh-speaking kingdoms emerged in southern Scotland. Gododdin was based in the Lothians and the Tweed Basin before it met the beginning of its end at Catterick in 600 and was overrun by Anglian war bands in the decades following. Rheged was just as fleeting and its history even more shadowy, but it was ruled by a great king, a war leader with something approaching a historical personality, someone generously praised by the bards and a man who met a tragic end at the hands of treachery.

Urien's name comes from Urbgen and is sometimes written so. It means 'Born in the City' and must be a reference to the fading townscape of Roman Carlisle. In the middle of the sixth century, the approximate period of Urien's birth, it still looked like a Roman town with many of its stone buildings surviving intact, its streets paved and at least one fountain fed by an aqueduct.

Urien's kingdom of Rheged stretched from Dunragit (originally Dunregate, the Fort of Rheged) near Stranraer in the west to Carlisle, its strategic focus and hinge, and it probably included Cumberland and Westmorland, now Cumbria. Cumber was the name coined by the Angles for the native British and it is cognate to Cymry, the Welsh word for the Welsh. Urien is thrice described as Lord of Echwydd, a term denoting a flow of water, a tidal current. This must be a reference to the Solway Firth and Urien may be seen as its ruler, the king of the Solway.

He raided far and wide for cattle and other booty but in the 590s his ambition was suddenly cut short. In an alliance with other native kings of southern Scotland and northern England, he led an attack on the growing power of Bernicia. Having pursued and trapped the Anglian war bands on the tidal island of Lindisfarne, the allied army was poised to strike a fatal blow when Urien was assassinated in his tent out of jealousy, at the instigation of Morcant, because his military skill and leadership surpassed that of all the other kings. Morcant Bwlc may have ruled the sub-kingdom of Calchvynydd on the Tweed at Kelso and perhaps he feared the domination of Rheged if the war against the Angles had been won at Lindisfarne.

After Urien's death, some time in the 590s, the great realm of the Solway began to fragment. His heir, Owain, the Son of Prophecy, was a far-famed war leader but his reign appears to have been short. By the time Oswy of Northumbria had married Urien's great-granddaughter in the 630s or 640s, his war bands had taken over Annandale and were pushing westwards beyond the Nith and into Galloway. Around the year 700 domination in Dumfriesshire was complete and the great Ruthwell Cross was raised, perhaps the most beautiful artefact of the age and the bearer of very early English poetry. The 'Dream of the Rood' tells the story of Christ's crucifixion. By 731, the Northumbrian version of Christianity reached as far as the ancient church of Ninian at Whithorn and it had become the focus of a new bishopric. Twenty years later King Eadberht was leading war bands into the Old Welsh-speaking sub-kingdom of Aeron, modern Ayrshire.

It seemed that, on three sides, Strathclyde was surrounded by the rising floodtide of English but it endured. One of Urien's allies in the fateful expedition to Lindisfarne was Rhydderch Hael who ruled on the Clyde at the end of the sixth century. His principal fortress was Alt Clut, the Rock of the Clyde, now known as Dumbarton Rock. Rising sheer out of the firth, it is mighty and impressive and not only did it command the sea road up and down the Clyde, the boats beached at its base could make their way inland up the River Leven and into Loch Lomond. Strathclyde warships could quickly be rowed into the heart of the mountains to the north so that they could effectively police the frontier with Dalriada and Pictland. It was marked emphatically at the head of Glen Falloch by a huge boulder still called the Clach nam Breatann, the 'Stone of the Britons'.

Strathclyde's most renowned saint, Kentigern, was a Briton but his origins lay to the east in the failing kingdom of the Gododdin. After spending time at Culross in Fife with his mentor, St Serf, according to his hagiographer, the young man founded a monastery on the banks of the Molendinar Burn, where it flows into the Clyde at Govan. Later in his exemplary life, Kentigern was also associated with the monks at Hoddom, perhaps in the early seventh century.

As the fortunes of the kingdoms surrounding Strathclyde ebbed and flowed, it resisted incursion, preserved its ancient native character and its kings continued to rule until well into the eleventh century.

Christianity appears to have come late to the people of Pictland. According to Bede, Nechtan, son of Derile, ruled from 706 to 724 and, at that time, wrote to Ceolfrith, abbot at the twin monasteries of Monkwearmouth and Jarrow (where Bede lived), asking for advice on how to set up a national church after the Northumbrian model. It may be that a number of factors were at play. Perhaps only the Pictish aristocracy had converted and Christianity was a patchy affair in Nechtan's kingdom, only adopted here and there. In 717, priests from Iona, possibly missionaries, were expelled from Pictland and it may be that one version of the Word of God was being replaced by another. Most of the very impressive Christian cross slabs were carved and set up after this time and the Pictish church seems to have become better organised and more assertive.

After he was defeated in a civil war in 729, Nechtan gave way to Oengus, another king who reigned for a long time. He had a Gaelic name but ruled from 729 to 761 without serious challenge. Oengus endowed a church at St Andrews, a foundation that grew to great eminence. The dynastic link between Dalriada and Pictland was confirmed when Constantine I began to rule in 789. He was of royal Argyll blood, a son of the Cenel nGabrain and, after 811, was listed as King of Dalriada. Scotland was slowly beginning to form but the process of unification was dealt a mighty blow during Constantine's long kingship. Here is the epoch-making entry for the year 793 in *The Anglo-Saxon Chronicle*:

> Terrible portents appeared in Northumbria, and miserably inflicted the inhabitants; these were exceptional flashes of lightning and fiery dragons were seen flying in the air, and soon after in the same year the harrying of the heathen miserably destroyed God's church on Lindisfarne by rapine and slaughter.

As if out of nowhere, the first Viking war bands had driven their longships up on to Lindisfarne's pebble beaches and a period of unparalleled ferocity began as Britain's shores were terrorised by the men known to monks as the Sons of Death. Over a long period, more than two centuries, they gradually became the fifth nation of Scotland and, if he had still lived, a people Bede would not have welcomed.

9

Of Mice and Men

THE SLEEK AND SWIFT longships of the Vikings were crucial to the reach of the early raiding parties and enabled surprise attacks to be made all along the British and Irish coastlines but they carried more than bands of ruthless and bloody-thirsty warriors. Somewhere in the corners and crannies of these open-decked dragon ships, another group of passengers were hidden. Mice came with the Vikings and, just like their human hosts, these tiny creatures colonised the Northern Isles, Caithness and parts of the Hebrides.

When researchers tested the mtDNA of house mice on Orkney, they found that it was quite different from that of their British cousins. Instead, it matched that of Norwegian mice very closely. And, more, native mice turned out to be related to Germany's rodent population. It seems likely that they made the journey to Britain across the watery plains of Doggerland.

And there the charm emphatically ends. What descended on Britain and Ireland after the 790s was shocking, savage and sustained, even by the violent standards of the times. At first, it was directed at easy and defenceless targets.

Mostly monastic in character, the early church was fascinated by the example of the Desert Fathers. These small groups of ascetics had fled the tumult of the world they found in the towns and cities of the Near East and also the occasional bout of persecution and they set up hermetic communities in the desert. Sometimes these were communities of one (the term monk derives from a Latin word meaning one). There these men could pray, fast and commune directly with God without interference. It was a life of concentrated spirituality, simplicity and privation, often deliberately induced. Mortification of the flesh was seen as a way of defeating appetitive desires and purifying the soul before God.

This approach to sanctity was adopted in the west and, in Britain and Ireland, the lack of deserts was compensated for by the endless wastes of the ocean. Monasteries were built in remote, sea-girt places like Iona and Lindisfarne and, in Gaelic, they became known as *diseartan*. It was their particular atmosphere, their lonely spirituality that made these communities perfect, highly vulnerable prey for the ferocious heathens who sailed their longships, the dragon ships known as *dreki*, across the North Sea. Monks began to look out anxiously over the horizon, praying that bad weather might protect them. One man wrote this in the margins of his gospel:

> The bitter wind is high tonight
> It lifts the white locks of the sea;
> In such wild winter storm no fright
> Of savage Viking troubles me.

The raid on Lindisfarne in 793 may have been the bloody work of splinter group from a larger fleet. There were attacks on the Western Isles in 794 and on the monastery of Columba at Iona in 795. The longships were dragged up the beaches below the church again in 802 and, when they returned yet again in 806, there was an appalling bloodbath with the slaughter of 68 monks. The shock rippled through the Christian church. Here are reactions from Alcuin of York, a scholar at the court of Charlemagne:

It has been nearly three hundred and fifty years that we and our fathers have lived in this most beautiful land. Never before has such a terror appeared in Britain and never was such a landing from the sea thought possible. It was terrible to behold the church of St Cuthbert spattered with the blood of Christian priests.

What appalled Christian communities almost as much as the hideous violence was the failure of the saints such as Cuthbert and Columba. They were thought to have the power to protect their followers, and if the Sons of Death could attack, ransack and kill at will, then that could mean only one thing. It was not that the saints were unable to shield their monks from the horror but rather that they refused to. Somehow Christian Britain and Ireland had incurred the wrath of God and his saints and their people, even their priests, were being punished by the fury of the Northmen.

After the slaughter of 806, Abbot Cellach made plans to abandon Iona. This was not done lightly. Around the monastery a ditch and bank had been dug to delineate the sacred precinct, the holy ground where Columba had walked. But the heathen Vikings were not deterred and such relics as remained had to be protected and a safe place for them found. Far from the sea, near the middle of Ireland and not on a navigable river or loch, Kells was chosen and most of the Ionan monks resettled there. Kells gave its name to the gorgeous illuminated gospel even though it seems that most of the work had been done on Iona.

Not all of the brothers followed Cellach to Ireland. Those who remained in the defiled monastery knew the dangers and it appears that some actively sought martyrdom at the hands of the heathens. In a secular life, Blathmac had been an aristocratic Irish warrior but his conversion to monasticism had been so profound that it seems he went to Iona precisely because it was raided so often. He wished to die for his faith, perhaps hoping to assuage the wrath of God with the ultimate sacrifice. He could not know how he would suffer.

An account of what happened in 825 was written by Walafrid Strabo, Abbot of Reichenau in southern Germany. Like many raiders wishing to use the element of surprise, the Vikings attacked the monastery at first light and broke into the abbey where Blathmac and his followers lay prostrate in prayer. In what must have been a terrible orgy of ferocity, they were all slaughtered before the altar except for their leader. Demanding to know where the monks had buried Columba's reliquary and other precious objects, the Vikings began to torture Blathmac. Probably using ponies, they took ropes attached to their harness and tied the ends to his arms and legs and, when he continued to refuse to give up the holy relics, the pious sacrifice was torn limb from limb.

The appalling fate of Blathmac was by no means unique. Amongst recorded atrocities – and there must have many that were not written about – the pagan ritual of blood-eagling was horrific. As a sacrifice to the war god Odin, victims were tied face-first to a post or pillar before a Viking marked the blood-eagle on his back. In 869, King Edmund of East Anglia suffered this dreadful death when his ribs were hacked from his spine and pulled outwards like an eagle's wings. Then his lungs were wrenched out and draped over his shoulders. On Orkney the Viking Earl Torf-Einarr ordered the same ritual in the 870s.

What drove this terror westwards, what they called 'westover-sea'? In Norway and over the rest of Scandinavia pressure had been building for some time on land and resources. Politics was concentrating and kingdoms forming. By the eighth century Danish kings had caused an immense earthwork to be dug across the neck of the Jutland Peninsula. The Danevirke marked the unmistakable southern frontier of an emerging kingdom. Across the Skagerrak, King Harald Finehair had completed the consolidation of the kingdom of Norway by the 880s. All of these tensions had the effect of driving ambitious men on to their longships to seek opportunity elsewhere.

After the first shocking flush of raiding, Vikings began to overwinter in Britain and Ireland and then settle. Fertile farmland had been at a premium on the steep sides of the fjords of Norway and, when settlers saw the green and flat fields of Orkney and Shetland, they moved aggressively to take them. Pictish natives were dispossessed, many probably slaughtered or driven away. Place-names track this process and the Northern Isles saw the introduction of Norse versions everywhere – *dalr* for 'a valley', *byr*, *stadir*, *setr* and *skali* which all mean 'a farm' and many others. These mutated into modern suffixes, often attached to personal names.

Settlement in the Northern Isles did not somehow lead to a softening of Viking activity or slake their thirst for blood. Far from settling down, they used Shetland and Orkney as bases for raiding in the south. Here is a very matter-of-fact account of a typical year in the life of Svein Asleifsson taken from the *Orkneyinga Saga*:

> In the spring he had more than enough to occupy him, with a great deal of seed to sow which he saw to carefully himself. Then when the job was done, he would go off plundering in the Hebrides and in Ireland on what he called his 'spring trip', then back home just after midsummer where he stayed till the corn fields had been reaped and the grain was safely in. After that he would go off raiding again, and never come back till the first month of winter was ended. This he used to call his 'autumn trip'.

Svein Asleifsson was described by the great Orcadian author Eric Linklater as the 'Ultimate Viking'. Although he nominally owed allegiance to the twelfth-century earls of Orkney, Svein appears to have acted completely independently. From his island fastness of Gairsay (and its drinking hall large enough for eighty men), which stands sentinel at the mouth of Wide Firth, the bay at the heart of the archipelago, he set sail for plunder each spring and autumn as detailed above. Ranging far down the western coasts of Britain, the Gairsay longships attacked Wales, laid siege to the island of Lundy in the Bristol Channel and took Svein to the Isle of Man to seek a bride so that he could seal a strategic alliance with a Manx sea lord. Eric Linklater was only echoing the *Orkneyinga Saga* of c.1200, for its composer hailed the Ultimate Viking as 'the greatest man in the Western Lands, either in olden times or present day'.

Ancestral DNA research suggests that when the last real Viking died in an attack on Dublin in 1171, his marker carried on – with great vigour. A relatively new sub-type of M17, S375, is now prominent in the North Isles of Orkney, on the five major islands north of Gairsay. It is also carried by 30 per cent of men with the surname of Gunn who have taken DNA tests. Tradition, genealogy and history all begin to come together to form a narrative. It may be that the prevalence of S375 on the islands of Rousay, Westray, Eday, Sanday and Stronsay is linked to the well-attested phenomenon of social selection, where powerful men in the past sired many children with different women. In that way their Y chromosome markers were spread widely and quickly, much faster than if they had remained monogamous. And few men were more powerful in twelfth-century Orkney than Svein Asleifsson.

The sort of brisk, well-organised raiding mentioned above was a continuing affliction for its victims and, in the sagas and poetry of the Vikings, it is described with evident relish. Even after they converted to Christianity, they never hesitated to sharpen their swords in the spring, push their longships out into the firths and hoist their sails for the south. When another Orcadian, the poet Bjorn Cripplehand, described an eleventh-century expedition, he was also celebrating centuries of raiding in the Western Isles:

> In Lewis Isle with fearful blaze
> The house-destroying fire plays;
> To hills and rocks the people fly
> Fearing all shelter but the sky.
> In Uist the king deep crimson made the lightning
>   of his glancing blade;
> The peasant lost his land and life
> Who dared bide the Norseman's strife.
>
> The hungry battle-birds were filled
> In Skye with blood of foemen killed,
> And wolves on Tiree's lonely shore
> Dyed red their hairy jaws in gore.
> The men of Mull were tired of flight;
> The Scottish foemen would not fight
> And many an island girl's wail
> Was heard as through the Isles we sail.

On Sanda's plain our shields they spy:
From Islay smoke rose heaven-high,
Whirling up from the flashing blaze
The king's men o'er the island raise
South of Kintyre the people fled
Scared by our swords in blood dyed red,
And our brave champion onwards goes
To meet in Man the Norsemen's foes.

In the light of all this violence and the glorification of it in Cripplehand's poem, it is surprising that many Scots hanker after Viking ancestry. Perhaps the attraction is the sense of adventure – a band of brothers setting sail, braving the mighty ocean and seeking their fortune. It is true that the Vikings were superb, skilled and daring sailors. In their beautifully designed but open-decked dragon ships, they sailed out of the sight of land and crossed the broad expanse of the North Sea to Britain. Even though Scandinavian shipwrights had perfected the design, with a keel deep enough to prevent capsize in a heavy sea, it still took tremendous courage to sail such distances without a compass.

The dragon ships could be very large, sometimes 80 ft long and 17 ft wide with a carrying capacity of 18 tons and a crew of up to 70. Their speed and manoeuvrability came from a combination of a square sail set to the wind and rows of oarsmen who could accelerate the ship very quickly. Guided by the steer board at the stern (the origin of starboard), the minimal draught needed for the keel meant that the longships could be rowed up even shallow rivers to reach targets far inland. Even if the draught was only a foot or two, a sea lord could lighten his ship by having all but two oarsmen and the steersman disembark. That way it could be moved upstream like a very large rowing boat. And, when Vikings attacked directly off the sea, as they did at Lindisfarne and Iona, where they might easily be seen by lookouts, the design of the ship allowed them to be rowed at speed very close to a beach before rasping against sand or shingle. Vikings used their dragon ships wherever it was possible and even where it was not. Between stretches of navigable water, crews would sometimes put the craft on rollers and drag it overland.

Made from wood sawn into long planks or strakes, they could be heavy – and certainly very much harder to move than curraghs. The strakes were overlapped, nailed and then lashed tight with pliable and tough spruce tree roots as strong as any rope before being caulked with horsehair and grease. Because the keel was usually made from the trunk of a single oak tree, the fabric of the whole ship was flexible and, when struck by big seas, the gunwales could twist a long way out of true but still remain watertight.

Navigation was not a precise art in the ninth century. For sea lords setting sail from Scandinavia, the eastern coastline of Britain and the Northern Isles presented a long target that was difficult to miss. Mist and fog were much feared, and when visibility was reduced distances had to be estimated by shifts at the oars or a floating log with a line attached to it. This was knotted at regular intervals and speed could be measured by dropping the log in the sea at the bow and working out by counting the knots as they unspooled the line how long it look for the stern to reach the log. The linguistic legacies are logbooks and knots instead of miles.

Sea lords watched the water and, when it was clear, they looked up at the sky. At night they could navigate by the relative positions of constellations and individual stars and in daytime they watched for seals inshore or for pelagic fish or birds. Sometimes they followed the whale roads to the north and many knew where the ocean currents ran in different places. These affected wave patterns and caused swells and eddies recognisable to seasoned sea lords. This sort of lore could shade into the mystical with the notion of the Mother Wave and how it could show the direction the sea was running. In Shetland, this is called the *moder-dye* – what sailors used for finding their way into safe harbour when a blanket fog had descended. They believed they could see different wave patterns as their boats edged closer to the land.

In bad weather, voyages must have been very hard. It would have been difficult for men to survive for long on the open decks of the *dreki* when they were constantly wet, constantly cold. These conditions may have been relatively rare. In the raiding centuries, there is evidence to suggest that there was a climatic optimum – a period when temperatures were higher and summers longer. Even so, the North Sea could be grey and inhospitable. Oarsmen began to slather fish oil on animal skins to make them waterproof and when the wind and rain blew across the bows, these first oilskins will have been welcome – despite the smell.

The reach and enterprise of Viking sea lords coincided with a marked increase in trading activity across Western Europe. The stability encouraged by the powerful Carolingian Empire and the new Muslim states of Spain and North Africa created opportunity and growing demand. From the Baltic and the far north, the Vikings brought amber, much-prized walrus ivory and eiderdown. These high-value and small-bulk cargoes were ideal for the longships and the larger, slower merchantmen called *knorrs*. But the most consistently lucrative and widespread trade carried on by the Vikings was slavery.

Dublin was founded in the 840s first as a *longphort*, a ship camp, and then it developed as a busy slave market. When sea lords raided, the more sober and business-like would not allow the younger and fitter people they captured to be slaughtered. Instead, they herded them on to the longships and took them to Dublin or elsewhere for sale. In 870, a huge fleet of Dublin Vikings, perhaps 200 ships, sailed into the Firth of Clyde and laid siege to Dumbarton Rock. This was an unusual strategy for raiders but there were many sheltering behind the palisades who would fetch a good price. When Dumbarton finally fell after three months, the greatest prize was not gold or silver but the Strathclyde royalty and nobility. These high-born captives would have commanded a premium. The Muslim states in particular set a high value on Christian slaves and some of the men will have been castrated for that market, adding great pain to their humiliation.

Women were widely traded – often for their looks as well as their skills – and at the slave market at Marseilles, an Anglo-Saxon girl, almost certainly captured by Viking raiders, was very fortunate. Balthild became the wife of Clovis II, the King of the Franks in the mid-seventh century and she spent her husband's money buying and then freeing slaves all over his realm.

The wide and somewhat jumbled distribution of mtDNA may owe something to the great revival of the slave trade (the Romans were perhaps the greatest slave masters of Europe – there were thought to be a million in Italy alone in the first century AD) by the Vikings. It appears to have focused on certain sections of British society. The Anglo-Saxon word for a native Briton was *wealh* (Welsh is derived from it) and it was also used to mean 'a slave'. At Corbridge by Hadrian's Wall, there was a slave market, well placed between the later Viking kingdom of York and the kingdoms in Scotland.

The discovery of both the pan-British Isles marker of S145 and the Irish and Scottish M222 in coastal Norway has suggested a remnant legacy of slaves shipped back to the Viking homeland. Even very small numbers of M284 have been detected. Although many Scots visited and even settled for long periods in Norway from the later Middle Ages onwards, it is quite possible that some of these S145 and M222 descendants are, in fact, the children of slaves. The British-specific J1b1 mtDNA group has also been found in coastal Norway and may be a female counterpart, an arrival less easily explained by post-medieval contacts with Scotland.

There is a fourth distinctly Irish subtype of the great S145 marker but, like the Pictish subgroup, it has yet to be identified with a single, slowly evolving marker. Instead geneticists rely on a particular signature of more quickly evolving markers to identify members of this group. It is concentrated in Munster and particularly in Counties Cork and Kerry. It is very rare in Scotland and has only been found in the Northern and Western Isles. This suggests that it is unlikely to have spread outwards, as M222 appears to have done, from Dalriada. Rather, it looks as if it was taken directly from south-west Ireland to north and west Scotland. A likely explanation would be that these lineages represent the descendants of Irish slaves taken north by the Vikings. This is supported by the fact that the major genetic lineage of the surname of Macaulay, the sons of Olaf, belongs to the group. It seems that some slaves contributed to the ancestral gene pool of the peripheral regions of Scotland.

In 839, a battle was fought in Strathearn that may have spelled the end of the Pictish kingdom south of the Mounth. A great force of Vikings slaughtered the Pictish nobility in such numbers that a power vacuum allowed Kenneth macAlpin to establish himself in Pictland in the aftermath. He may not have been the first Dalriadan king to rule east of Drumalban but all Scottish kings are numbered from him and, with his accession, a process of unification did appear to begin in earnest. With the near-fatal weakening of Strathclyde after the siege of Dumbarton Rock in 870, Vikings had a significant role in the making of Scotland.

By the later ninth century, the Vikings had settled and colonised the Northern Isles, Caithness, the Hebrides and the Atlantic coasts of the mainland. Shetland comes from the Norse name, Hjaltland and Caithness from Katanes, but Orkney did retain its original Pictish name. The relative density of settlement can only be guessed at but it must be indicative that in the Northern Isles the Norse language replaced Pictish, while in the Hebrides Gaelic began to reassert itself in the tenth and eleventh centuries.

The genetic inheritance of this period of great turmoil in Scotland's history is fascinating. There are powerful and persistent links between Norway and the Northern Isles. In Orkney, 20 per cent of men carry the Y chromosome marker M17 and its frequency in Norway is 30 per cent. It is much rarer in the south and west of Scotland and England, reaching about 4 per cent, and it looked a likely candidate for a Norse or Viking marker. If it was, then the Y chromosomes of men with distinctive Orcadian surnames should contrast with those with Scottish Orcadian names. The Y chromosome is in most cases inherited with a surname.

If the former represent a Norse legacy, then they should carry a much more pronounced weighting of M17. Men whose lineages preserved the names of Flett, Foubister, Clouston, Isbister, Rendall and others were compared to holders of the likes of Spence,

Sutherland, Sinclair and Garioch, all of them old names but probably mainland Scottish ones. It turned out that the M17 marker was very much more heavily concentrated in the group of old Orcadian surnames with 75 per cent carrying this or another signature predicted to be Norse. They were, beyond doubt, the descendants of the Vikings who came to settle after the 840s. Perhaps the remaining 25 per cent represent the survival of even older lineages, a remnant of the Pictish society displaced at that time, or otherwise simply represent the effects of illegitimacies and surname adoptions over the last 30 generations.

In Shetland, many surnames were passed on as patronymics until about 1800, like the Mac-names of Highlanders. This custom, shared between Highlanders and the Norse, marked out a group with a higher likelihood of Norse ancestry as the Norse markers were well established with older families. Men with names like Williamson, Robertson, Matthewson, Thomson and Jamieson were known to carry the marker. Evidently Scottish Christian names were fashionable around the time these patronymics were fixed as surnames but the Norse bloodlines carried on into the future.

More recent Scottish names on Shetland did not have the M17 marker. In fact, researchers believe that up to half of all male lineages in the Northern Isles are Norse in origin while the proportion in the Hebrides is about one third and in Caithness a quarter. As results were gathered further south, there was a very significant drop-off and no Norse signal at all in north-east and south-east of Scotland.

M17 was the first of a number of markers of Norse Viking ancestry in the British Isles to be discovered. S68 and S182 are smaller groups within the great M269 group and they appear to originate in Scandinavia and are mostly limited to the Northern and Western Isles of Scotland. S142 is more complex. It reaches its highest frequencies in Scandinavia and is common in Denmark but quickly drops off further to the south. In Lewis, there are many men with this particular marker, some in a group known as the Ultra-Norse – clearly one with Viking provenance.

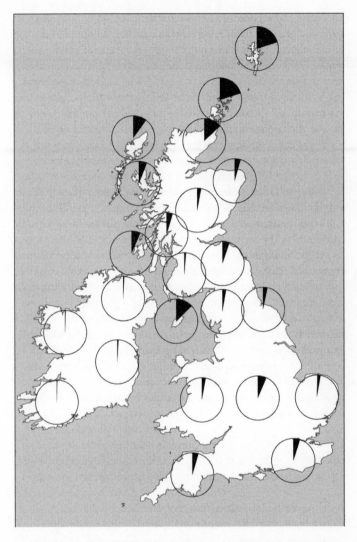

The frequencies of the M17 Y chromosome group are shown across the British Isles using pie charts. Up to 3,000 samples were used to create this map.

Vikings raided all over Western Europe and also moved eastwards to found the remarkable state of Kievan Rus – and, indeed, the name of Russia itself may have a Norse origin, a reference to the red hair of the incomers. They reached as far south as Constantinople, which they called Micklegarth, and formed a feared imperial guard called the Varangians. But perhaps their most dynamic colony was Normandy in northern France, the land of the Northmen, and, in the eleventh century, its vigour would begin to affect Scotland.

Meanwhile Constantine II had become King of Scots in 900, and through a long reign of at least 40 years (chroniclers disagree about exactly how many) he laid many of the foundations of the kingdom of Alba or, in English, Scotland. Constantine's armies defeated forces of invading Scandinavians in Strathearn in 904 and again at Corbridge in 918. With Bishop Cellach, he established something approaching the structure of a national church with its principal see at the shrine of St Andrew. Despite his defeat at the hands of the English king Athelstan at Brunanburh in 937, Constantine was able to resign his throne and retire to monastic life having achieved expansion and a degree of stability.

Dynastic rivalries crackled in mainland Scotland for the next century or so while the Norse earls of Orkney consolidated their northern possessions. By the time of Malcolm Canmore in the mid-eleventh century, the kingdom of Strathclyde had finally faded and most of Scotland had coalesced within its familiar boundaries. Only in the far north, the ancient land of the Vikings, did the writ of Scottish kings not run.

Conquests and Armadas

❈

B Y THE END OF THE first millennium AD, the age of the great migrations to Scotland was over. Elites would come and go, small ethnic groups would arrive, especially in the nineteenth and twentieth centuries, but there were to be no more large groups of new people. As the five nations of Scotland began to settle, their languages contended for wider adoption with one eventually coming to dominate. The others died away and failed, their ways of describing Scotland almost lost. In the Middle Ages and increasingly in the modern period, surnames became common and, with the keeping of parish records and the introduction of the universal census, the lives and lineages of ordinary people began to emerge from the background. Scots came to understand a great deal more about who they were and where they came from.

The long reign of Malcolm III Canmore (1058 to 1093) saw his dynasty consolidate and his descendants were to rule over Scotland until 1286 and the accidental death of Alexander III. Canmore is usually inadequately translated as 'Bighead', as though it was a physical characteristic. In fact, it means 'Great Chief' and it may be a memory of how contemporaries addressed him. The mac-Malcolms were indirectly descended from the macAlpins and both were scions of Dalriada and the Gaelic west. Because most Scots – indeed, most British people – are overwhelmingly monoglot with only English at their command, we often make an unconscious assumption that such blinkered attitudes have always prevailed. We are, after all, an island race and rarely come into contact with people who do not speak English. Also, with the viral wildfire spread of the Internet, it seems that soon the whole world will require to understand our language.

In the past, it was very different. In Malcolm Canmore's Scotland, there were six languages in common currency and probably many more local dialects babbling behind them. In the Northern Isles, Norse was spoken. In the Hebrides, the Atlantic west and Galloway, Gaelic dominated. Pictish is likely to have clung on along the North Sea coasts and perhaps in some more isolated inland areas. English was heard in the Lothians and the Borders. In the upper Clyde Valley and the central Southern Uplands at least, Old Welsh speech communities still existed. The sixth language was, of course, Latin which was used by the church and also for literature and for the keeping of secular records.

Many eleventh-century Scots will have been native speakers of one language, been able to get by in others and perhaps understand the gist of still more. People who lived on the boundaries of speech communities will almost certainly have been bilingual and aristocrats who owned and managed more than one property may have needed more than one language and, if they were literate, will have found it very useful to know what Latin charters meant. If goods and money were at stake, comprehension followed very quickly. To complicate further an already complex picture, another language would come to Scotland in the eleventh and twelfth centuries. After their startling successes in 1066, the Normans brought French to England and eventually to Scotland. It was to be the prime language of power for a long time.

Malcolm III Canmore married a Hungarian-born princess. She was, in reality, the daughter of Edward, the son of the English king Edmund Ironside. He had been exiled to Hungary by Cnut, the immensely ambitious Scandinavian ruler who incorporated England into his North Sea empire. Margaret spoke Hungarian and Early English, understood Latin (for she was very devout and was canonised in 1250) and when she arrived in Scotland in 1070 she would have had to learn to speak the wholly different language of the court, Gaelic.

DNA markers and languages are often linked – certainly in the case of Norse – but as the linguistic pattern of Scotland shifted and changed in the medieval period the picture became very much more blurred, especially at the edges.

The two indigenous languages in decline were Pictish and Old Welsh and this was very much a function of political failure. The kingdom of Strathclyde had extended south into Cumbria in the tenth century but its wide territory was finally absorbed forcibly into the realm of Scotland by Canmore in 1070. Nevertheless, the institutions of the Old Welsh speech community did not immediately wither, strongly suggesting that the language too had some resilience. A twelfth-century legal document, the *Leges inter Brettos et Scottos* (the Laws between the Brets and the Scots), was tidied up and updated during the reign of Canmore's son, David I (1124 to 1153). Before his unexpected accession, he had been styled Prince of Cumbria and in fact remained closely linked to the area, dying at his castle at Carlisle in 1153. More than his father, he would have been aware of the working of a different, Old Welsh-speaking culture in the old kingdom of Strathclyde. As an efficient, modernising monarch, it was David's instinct to see a proper legal framework bridging the Scots (that is, the people north of the Forth) and some of those south of that divide.

These laws talked of a Celtic society, one where the Brets, or Britons, saw families as a basic legal unit. They were responsible for any member who had committed a crime and bound to compensate a victim's family. The rights of women were guaranteed and marriages seen in a pre-Christian light as contracts between families rather than an ecclesiastical rite. If a husband was adulterous, a wife could divorce without a loss of property or face. Old Welsh words continued to be used for penalties for murder – the payment to a victim's family was called *galanas* – and, less seriously, that for insults was *sarhad* or *sarhaed*. There existed a clear tariff of recompense with rates listed for a king downwards and it was often paid in the classic Celtic currency of cattle.

After 1070, the culture and language of the Brets appear to have been in slow retreat, despite legal buttresses. What often happened is that more isolated, upland communities were the last to lose the old ways and speech. This process is difficult to date but place-names and their change or survival can offer hints of what happened and when. The scholar Tim Clarkson has made some recent and telling observations about the ancient Old Welsh-speaking territory of Goddeu. It meant simply 'The Trees' or 'The Forest' and it lay across the hill country where the Clyde and the Tweed both rise.

This area survived as a distinct entity into the modern period and the eastern part is still known as The Ettrick Forest, wild land used by Scots kings as a huge hunting reserve. But Goddeu was much older – the bards sang of it in the sixth century and Gurcyon Goddeu was said to have been the wife of Catraut Calchvynydd, a sub-king who fought in the Gododdin host at Catterick. The double d in Goddeu is pronounced in the Welsh style like the *th* in 'them' or 'they' and it appears to have survived in Cadzow, the old name for the town of Hamilton. Cadzow is sometimes pronounced Cadyow and it survives in Kilncadzow, an upland village whose name probably meant something like 'the Chapel of Cadzow'. In a slight variant, locals call it Kilcaidie. The confusing z in Cadzow came about through a much-repeated scribal error when the medieval letter of *yogh* was replaced by it. And so it may be that Cadzow is a more recent rendering of Goddeu. Certainly upper Clydesdale is speckled with Old Welsh names. Near Kilncadzow are Carluke, 'the Fort in the Marsh', and Carnwath, which is very suggestive of Goddeu for it means 'the Cairn of the Wood. And the *caer-* prefix reappears in Carstairs, the second element of which is probably a reference to a local stream.

Cadzow, modern Hamilton, was a focus of royal power for many centuries and its use by both earlier Strathclyde kings and David I shows a long continuity for a place of authority. Eventually the Dukes of Hamilton built a castle and then a palace before conferring their own name on Cadzow. By that time, the Old Welsh language had been long fled. And, in a curious, confused memory, the Dukes' residence at Chatelherault was the last in a line of hunting lodges for what is still known as the Forest of Cadzow.

The great toponymy scholar W. J. Watson produced his masterwork, the definitive *The History of the Celtic Place-Names of Scotland,* in 1926. It is a mine full of gems, and because Watson grew up in Easter Ross in the mid-nineteenth century on the border between Gaelic and English speech communities he had a wonderfully tuned ear for names and words and how they changed over time. In an unexpectedly personal passage in his great work, Watson described the process of language shift at first hand, something he witnessed in his own lifetime. It is a process that took place all over Scotland in the last millennium and what Watson described in Easter Ross could have happened in a very similar way between Old Welsh and English in Upper Clydesdale in the Middle Ages or between Pictish and Gaelic in Perthshire before that. It is a passage worth quoting at length:

> The process that took place may be illustrated by what went on and is still going on in the course of the change from Gaelic to English. When the Gaelic speakers began to take to speaking English, they made up their deficiencies in that language by using Gaelic words freely instead of the English terms with which they were not yet familiar. This was a custom with which I was very familiar in my own native district a good many years ago. At that time the English of Easter Ross was full of Gaelic words for which we had no handy English equivalents. We said, for instance, 'I have a meanmhainn in my nose', i.e. a premonitory tickling sensation; a child who would not eat his porridge was told 'you'll be a taidhbhse', i.e. literally, 'a phantom'; of a pithless man it was said there was 'nothing in him but a blianach', i.e. 'a meagre creature'; and so on almost AD infinit um As time went on the Gaelic terms became fewer; the more proficient in English were inclined to make merry over those who were less proficient. The man who went to Invergordon and demanded in a shop some pounds of 'beef-uain' – i.e. 'beef of lamb' – instead of mutton, was known as 'beef-uain' ever afterwards; another worthy man earned the name of 'twenty-ten' because he used that expression for the more regular 'thirty'. At the present day the number of Gaelic terms in the English of that district is much less than it was forty years ago, but some survive still. Nowadays the influence of the public school [Watson means the state school here] and the newspaper is decisive against their becoming permanent, but in older times, when

conditions were more favourable, some of these Gaelic words became so firmly established that they are still current in the Lowlands Scots of the districts that were once Gaelic-speaking. The Garisch classic 'Johnny Gibb of Gushetneuk' contains several such survivors, e.g. 'ablach', 'bourach', 'clossach'; in Fife they still use 'carr-haun' for 'left hand', the first part being the Gaelic 'cearr', left. In Galloway corn was at one time separated from the husks by rubbing it with the bare feet: this they call 'Lomeing of the corne'; here 'lomeing' is Gaelic 'lomadh', 'stripping'. These examples may be sufficient to illustrate what takes place. The process is, of course, not peculiar to Scotland: it is well known in Ireland, and something of the same nature must have happened wherever one language has been displaced by another. From one point of view the terms which survive from the earlier language maybe said to be borrowed into the later, but they are not really loans in the ordinary sense. What was borrowed in this case, to put it somewhat paradoxically, was not the Gaelic but the whole of the English which the people gradually came to use.

The few British [Old Welsh] terms which remain in Gaelic are to be accounted for in the same way, not as loans, but as survivals of the older speech. Their survival was due doubtless to a certain fitness which may not have been the same in all cases. One condition which must have been effective was the presence or absence of Gaelic of a handy synonym conveying the same shade of meaning. Thus British 'obar', 'confluence', had no chance against Gaelic 'inbhear', for both had exactly the same meaning. On the other hand, 'preas', a copse, was a useful general terms for which it was difficult to find an exact Gaelic equivalent; in the east 'tom' has its British sense of 'rounded hill'.

With the patronage of the macMalcolm court behind it, Gaelic advanced into the east at least in terms of place-names and the way in which new owners changed them. Deep in the heart of Pictish Fife, two neighbouring villages show this clearly. Balmalcolm is unambiguous – it was renamed from the lost original to become Baile Maol Chaluim, 'the farm of Malcolm'. The Christian name is indeed Christian for its literal meaning is 'the Servant of Columba', usually a name for a monk or priest, and it may, in this case, have denoted church land. Further east along the A914 lies Pitlessie. It preserves the Pictish *pit-* prefix, meaning 'a portion of land' or 'a farm', and it is attached to what was probably a Pictish proper name. Over the rest of Fife, there are similar juxtapositions recording a patchwork of settlement and ownership – Pitscottie and Baldinnie are neighbours south-east of Cupar, and Pittenweem, 'the farm by the cave', is on the coast not far from Kilnconquhar, 'the church of Conchobar', a little-known Irish saint.

Malcolm Canmore's youngest son David was raised at the court of the Norman-French King of England, Henry I, where he undoubtedly became fluent in French. He was known there as David fitzMalcolm, a Norman contraction of *fils de* meaning 'son of'. At the Scottish court, he was known as David macMalcolm. On his unexpected accession, the new king brought his French-speaking friends north as a means of modernising the Celtic realm of his forebears. Large landholdings went to Walter fitzAlan in Renfrewshire who later adopted the name Steward from the office he and his family were appointed to at the royal court. They themselves became, in time, a famous dynasty. Annandale went to Robert de Brus from Brix, his lordship in the Cotentin peninsula in Normandy. His descendants too would become kings.

David I also imported communities of reformed orders of monks from France and England and their abbeys grew quickly into large corporations, exploiting the countryside as well as supplying a royal civil service. Towns were founded, and by 1153 there were 15 royal burghs holding regular markets, some of them minting coins and all of them providing revenue to the crown. It was a revolution of dizzying rapidity brought about by a modern monarch intent on bringing a backward nation up to date.

The ancient kingdom of Moray had intruded into the line of macMalcolm kings in the shape of Macbeth and it continued to support other claimants. Beyond it lay the Norse earldom of Orkney. By the early twelfth century, it had close ties to the kings of Norway. There seemed more than a possibility that trouble would come from the north. David I determined to establish his authority at least in Moray – much larger than the modern county, it compassed most of the Moray Firth shore not belonging to the Orkney earldom – and he sent some of his more rugged and ambitious Norman-French friends to underline it. Frasers, Grants, Menzies, Gordons, Oliphants, Hays and Chisholms were all originally relatively recent incomers – no matter how quintessentially Scottish their names sound now.

Just as the Frasers got their name from the lordship of Frezeliere in Anjou, the Grants were probably originally Le Grand and the Menzies (Manners in England) were from Mesnieres in France. The MacIvers, the MacSweens, the MacAulays and the MacLeods are also relatively recent arrivals. These clans all have Norse anteced-ents, being, respectively, the sons of Ivar, Svein, Olf and Ljot. But their DNA markers are sometimes surprising since many men who were natives or had other origins took the name of their clan chief so that they might enjoy protection and certain customary rights.

Preliminary indications of the ancestry of the Norman families who came to Scotland are very interesting. Given that the Duchy of Normandy was founded by Vikings, famously by Gongu-Hrolf, it might be assumed that Norman and Norse descent could be mixed and difficult to disentangle. But in fact a brief survey of several Norman-French surnames in Scotland has shown up little or no M17 and it looks as though these people brought French bloodlines to Scotland rather than more Vikings.

Nevertheless, there are some wonderful discoveries to be made about lineages amongst the northern clans. Clan MacLeod tradition-ally recognised their Norse ancestry and an analysis of their DNA is rewarding. From a sample of 45 Macleod Y chromosomes, almost half – 47 per cent – clearly show social selection at work in that they descend from one individual. If this statistic is projected amongst the total number of MacLeods, it means that almost 10,000 men are descended from this individual. Amongst the remaining 53 per cent, researchers have found only nine other lineages present, showing that MacLeod men married women who were unfailingly faithful to them!

However, the MacLeods do not carry the M17 marker group. Theirs is a recently discovered subgroup labelled S68. It is found in Lewis, Harris and Skye, core Macleod territory, but also in Orkney, Shetland and Norway, with a few examples in Sweden. Despite extensive screening, S68 is very specifically located, showing up only once in the east of Scotland and once in England. This is a classic pattern for a Viking marker in Britain but one much rarer than M17. MacLeods determinedly claim descent form a common name-father, a Norse aristocrat called Ljot, a relative of Olaf, King of Man. They are probably right to continue to claim that – science, for once, is supporting tradition.

To the south of the lands of the MacLeods lay the wide lands of Clan Donald. Their name-father was the first Lord of the Isles, Somerled, and, once again, social selection counts 50,000 men with the name MacDonald or its variants as his direct descendants. There is accurate data available from a large sample of 164 MacDonald Y chromosomes and it contains a fascinating twist on tradition. Somerled was known to chroniclers as 'Somerled the Viking' and it turns out that the large number in the sample descended from him – 23 per cent – carry a specific signature type within the Norse subgroup of M17. Somerled's own ancestors did indeed originate in Scandinavia. And the tradition lives on for Clan Donald have genotyped the chiefs of their various clan branches and they all carry the old Viking's marker.

Another large lineage cluster in the MacDonald sample has a very different origin. Around 12 per cent carry the classic R1bPict marker and it may be that they are descended from a powerful individual whose identity is now lost but who chose to join with Clan Donald and adopt the name. There are two mainland branches – MacDonell of Glengarry and Clan Ranald – and both have chiefs with the Somerled marker, but their followers may well be Pictish.

Clan MacGregor has a colourful record with men like Rob Roy and Sir Gregor MacGregor making slightly disreputable but dashing marks on history. Their clan lands were in Perthshire, on the eastern slopes of the Drumalban Mountains, firmly in the ancient domain of the Picts. From a sample of 144 MacGregor Y chromosomes, a large proportion – 53 per cent – clearly descend from one individual. The clan motto is *'S Rioghal Mo Dhream*, 'My Race Is Royal'. They claim lineage from Alpin, the ancestor of Kenneth MacAlpin, king of Dalriada and Pictland in the mid-ninth century. The problem with the tradition is that MacAlpin's DNA was almost certainly Irish/Celtic and that of the 53 per cent of the MacGregors who share a common ancestor is not. They all carry S145-Pict. Whoever

Gregor was, he is unlikely to have been a Dalriadan. And, with 53 per cent of the total sample being Pictish compared with only 7 per cent of the Scottish population, the clan is emphatically Pictish and possibly descended from royalty. But perhaps not the royalty most MacGregors have in mind.

What this brief tour around the sample statistics shows is that the Highland clans were not a homogenous group. They fall into at least four subgroups and they reflect part of the general genetic pattern of Scotland – the Norse group in the north, the Norman-French in the old kingdom of Moray, the Pictish in the centre and the Irish/Celtic group in the south-west. Social selection has clearly played a very influential role and the translation of clan as *clann*, 'children of', was literally true in many cases.

Galloway is as far from the Highlands as it is possible to be in Scotland and yet it was home to a Gaelic-speaking culture until the early modern period. In fact, the territory of Greater Galloway that stretched east to Annandale and north to include Carrick may be seen as a palimpsest of our linguistic and cultural history, a mirror to what happened in perhaps more familiar parts of the country.

The most westerly peninsula, the Rinns of Galloway, lies close to Ireland and, at the same time as Dalriada was emerging, Gaelic was certainly spoken there. The kingdom of Rheged understood itself in Old Welsh and when it faded and died the English-speaking Bernicians pushed westwards to establish an Episcopal see at Whithorn and colonise fertile costal areas. Even Pictish became part of the mix – by mistake. The Bernicians may have believed the Gaelic speakers of Galloway to have been Picts because the first two bishops took symbolic names. Peohthelm means 'Leader of the Picts' and Peohtwine 'Friend of the Picts'. In the late ninth and early tenth century, the kaleidoscope was twisted once more when some of the Celto-Norse peoples of the Hebrides migrated south. Because they spoke Gaelic but were descended from Vikings, they became known as the Gall-Gaidheal and they gave their name to Galloway. It means 'the Land of the Stranger Gaels'.

In a brilliant analysis, the toponymy scholar W. F. H. Nicolaisen has traced the first arrival of early Gaelic and the people who spoke it in Galloway. Noting that the place-name *sliabh*, meaning 'a hill', was found in the Rinns peninsula in great density and also on Islay, Arran and other districts of Dalriada but very rarely elsewhere in Scotland, he began to detect hints of population and language shift. The forms of *sliabh* used were themselves early and certainly predated the arrival of later dialects of Gaelic in the longships of the Stranger Gaels. The migration of Irish speakers in the period around 500 is not attested anywhere in literary, archaeological or even traditional sources, and when Nicolaisen revealed this important new evidence, found solely through an examination of place-names, there is a justifiably triumphant note in his writing.

Contact or conflict between the Gaelic speakers of the Rinns and the native Old Welsh communities is not recorded either but it seems that the basic British stratum of society developed undisturbed. Again place-names are important – the Old Welsh *tref*-prefix, meaning 'a farmstead', is seen all over the map of Galloway, at Troqueer, Terregles, Threave and elsewhere. Populations were organised in kindred groups and, long into the historic period, there existed a role for a man known as the *pencenedl*, the 'chief of the kindred'. Land was divided into administrative units that resembled the cantreds of Wales.

As late as the twelfth century Old Welsh was still spoken in Nithsdale and Annandale. There, a man with the Gaelic name of Gille Cuithbrecht (servant of St Cuthbert, the famous Bernician saint) encapsulated the cultural diversity of Galloway. His nickname was apparently 'Bretnach', a Briton, someone who also spoke Old Welsh.

When the Bernicians began to arrive in the eighth century, their settlements concentrated in three clusters. Once again, the density of English place-names show them taking over good land around Dalbeattie, Kirkcudbright and Whithorn. It is not always clear which group was dominant and no doubt the pattern varied but two words still current in Galloway dialect in the nineteenth century offer clues. Both *gossock* and *kreenie* meant 'a servant' or 'a lower class sort of person' – *gossock* derives from *gwasog*, an Old Welsh word for 'someone of servile status', while *kreenie*, almost a term of abuse, is from *cruithnich*, 'a Gaelic speaker'. There appears to have been no equivalent in early English and perhaps that too is eloquent.

Around the year 900, the Gall-Gaidheal began to arrive in

Galloway and the spread of Gaelic began to gather pace. It may have reached as far east as the Annan for, beyond it, Gaelic place-names quickly thin out.

The chronicler Richard of Hexham repeated Bede's assertion that the Gallovidians were Picts and, despite the early ministrations of their Pict-friendly bishops, they acquired a reputation for barbarous ferocity in the Middle Ages. This may have been an effect of the Gall-Gaidheal Viking heritage. With such feared warriors at their command, the Lords of Galloway emerged as a powerful political force. At the beginning of the eleventh century, Suibne macCinaeda was asserting himself as King of the Gall-Gaidheal, and a century later his descendant, Fergus, styled himself a little more grandly as King of Galloway.

Such pretensions, even on the Celtic fringes of his kingdom, were anathema to David I and he began to hedge Galloway with grants of land to Norman-French (by this time Anglo-Norman) families loyal to him. The Bruces settled in Annandale and Carrick and others were given estates elsewhere. When Fergus died in 1161, his sons, Uchtred and Gille Brigte (sometimes anglicised as Gilbert), disputed the succession and in a brutal Celtic tradition the loser was blinded and mutilated. Gille Brigte died soon after, the victim of ancient beliefs that kings without sight and physical imperfections could never rule. The same horror had been visited on King Donald II Ban in 1099 when he was captured by King Edgar.

Several powerful Gall-Gaidheal families had developed into clans although they did not use the term. Instead, they harked back to an earlier in-migration of Gaelic speakers. Those attached to the Kennedys, the Maclellans, MacDowells and Mac-Culloughs called themselves 'kenelmen' after *cenel*, the old word for 'a kindred', and a chief was *kenkynol*, from *ceann-cinneal*.

Agriculture began to be described in Gaelic and place-names remember great precision. The word for an *airidh* or shieling is embedded in Airies and Airieholland, while Talnoltrie is from *talamh an oltraigh*, 'an inbye field mucked by beasts in the wintertime'. There are many other equally detailed examples.

By 1234, the great lordship had failed. While Alan of Galloway was wealthy and widely landed (and had not irritated the king by calling himself one), he could not father a male heir. Alexander II of Scotland would not allow his illegitimate son Thomas to succeed. The line was broken and the old kingdom began to dissolve. But in the west and in Carrick the Kennedys remained significant. It seems that Suibne macCinaeda bore a version of their name and it appears to derive from the Old Welsh title *Cunedda*, simply meaning 'Good Leader'. A renowned general who led an expedition from Scotland to Wales in the fifth century had the same title and in the medieval period he was still celebrated by the bards. To scions of Clan Kennedy who knew their own history, it will have seemed doubly appropriate when the most famous son of the name, John F. Kennedy, was elected in 1960 to become the most powerful individual in the world.

Galloway's Gaelic was more durable than its kings or lords. It was still vigorous enough to be attacked by the poet William Dunbar in the sixteenth century – he called it Irish and claimed an English speaker could fart more eloquently. But, by the seventeenth, it was dying. A Margaret MacMurray was one of the last native speakers but the speech of the Gall-Gaidheal may have lingered a little longer in the words of Alexander Murray. He was an eminent, self-taught philologist and it is thought that he learned Galloway Gaelic from his father, an upland shepherd from the Stewartry of Kirkcudbright.

Surnames begin to appear in the historical records of medieval Scotland during the reign of David I. His aristocratic Anglo-Norman friends were the first to have and keep them. Most are location-specific and refer to lordships in northern France – de Brus came from Brix and became Bruce and de Ridel did not change much when it was written as Riddell. Others did not travel so far. Grahams came originally from Grantham in Lincolnshire. In addition to those names that migrated north ultimately to morph into Highland clans, the most common in Scotland of Anglo-Norman origin are Bissett, Boyle, Colville, Corbett and Kinnear.

In 1296, the Hammer of the Scots, Edward I of England, dismissed a Scottish king with a French name and land in Galloway, the hapless John Baliol, and demanded that prominent Scots swear homage to him. Their names are recorded in the Ragman Rolls. It lists a mixture of surnames – some are place-related, others occupational and many are patronymics. The form of the latter still, of course, survives in Celtic Britain in the shape of mac-names like MacLeod, MacCullough and MacDonald. They were modified in Wales by officials who did not have the language from names such as Ap Rhys to Price or Ap Hywell to Powell. Occasionally, an ancient ap-name survives in a Scottish medieval source and in Dumfriesshire there is the splendid MacRath ap Molegan. He seems to carry both a Gaelic and an Old Welsh version of 'son of' in his excellent name.

One of the most fascinating in Scotland is Galbraith, for it seems also to span two cultures and two languages. It is from *Gall Breathnach* or 'Stranger Briton' and the family came from Lennox, a frontier area between Strathclyde and Dalriada. Clearly the Galbraiths did not call themselves Galbraith but the name stuck nevertheless.

To outsiders, the notion of a Gallovidian or Highland clan sharing a common surname may have appeared confusing. How did 10,000 men all called MacDonald distinguish themselves one from another? It was done in two ways. Each man was usually able to recite his genealogy and that, by itself, could be sufficient. In 1585, a man from the island of Lismore was recorded at length as John Roy macIain VcEwin VcDougall VcEan or, in modern Gaelic, Iain Rudh, macIain, mhic Eoghainn, mhicDhugaill, mhic Iain and, in English, Red John, son of John, grandson of Ewan, son of Dougal, son of John – *mhic* means 'grandson'. There was no mistaking who this man was and many Gaels still glory in reciting long and proud genealogies. Less time consuming were 'to-names' or nicknames and Red John clearly had one of those too – probably he was a man with red hair.

MOST COMMON SURNAMES IN SCOTLAND
The percentages are reckoned as a proportion of all surnames in Scotland

1. Smith (1.2%), (occupational name)
2. Brown (0.94%), (nickname)
3. Wilson (0.89%), (patronym)
4. Robertson (0.78%), (patronym)
5. Thomson (0.78%), (patronym)
6. Campbell (0.77%), (nickname)

7. Stewart (0.73%), (occupational name)
8. Anderson (0.70%), (patronym)
9. Scott (0.55%), (ethnic name)
10. Murray (0.53%), (territorial name)
11. MacDonald (0.52%), (patronym)
12. Reid (0.52%), (nickname)
13. Taylor (0.49%), (occupational name)
14. Clark (0.47%), (occupational name)
15. Ross (0.43%), (territorial name)
16. Young (0.42%), (nickname)
17. Mitchell (0.41%), (patronym/nickname)
18. Watson (0.41%), (patronym)
19. Paterson (0.40%), (patronym)
20. Morrison (0.40%), (patronym)

The common surnames of Little, White, Brown and many others were originally nicknames. One of the most common in Scotland is Campbell, accounting for 0.77 per cent of the population. It comes from a Gaelic phrase *cam beul* or 'twisted mouth'. Cameron is also not immediately obvious and it is from *cam sron*, 'crooked nose'.

Patronyms were not exclusive to Highlanders and Gallovidians. Wilson, Watson, Thomson, Anderson, Morrison and Paterson all appear in the list of the 20 most common Scottish surnames compiled in 2001. Most of them are Lowland in origin. Other names, like Martin, dropped the 'son' element and some reduced it to a terminal 's' as in Adams.

While Anglo-Norman names were attached to landowners and their land, they did also introduce a series of diminutive suffixes into Scotland – for example, -el, -ett, -on, -in, -cock and others. Paton is a diminutive of Patrick and Adkins of Adam.

A final group of surnames was occupational. Baxter is the Scots word for a baker and the likes of Mason, Smith and Taylor are self-explanatory. Names also changed with the circumstances of those who bore them. Highlanders sometimes dropped the mac, probably in the face of prejudice, when they came south so that, for example, MacCowan became Cowan or in a different way, MacIlroy was squeezed into Milroy. Simple anglicisation converted MacDonald to Donaldson, MacIain to Johnson and MacDonnchaidh to Duncanson or Duncan.

This last group can be unhelpful to geneticists. Surnames are unique cultural markers of shared ancestry. In the absence of change, surname adoption and illegitimacy, they are inherited together with the Y chromosome. Men tend to keep their names and women change them. English research has revealed a remarkably high degree of shared ancestry within a surname and that increases dramatically the rarer the name. All Attenboroughs are related, for example, and so are Herricks, Striblings, Swindlehursts and Haythornthwaites. On average, around 20 per cent of the bearers of a common name are related and may be descended from a single name-father, rising to 40 per cent for rarer names. Clearly, rare names skew this percentage and with the prevalence of occupational names and nicknames like Smith and Brown, the percentage will be very much lower. There were hammer-wielding Smiths, needle-pushing Taylors and dough-pummelling Baxters or Bakers all over Britain and Ireland who are now entirely unrelated but, nevertheless, surnames matter.

Identity mattered a great deal in the twelfth century in Scotland but in a different way. As a widely landed English magnate holding the earldom of Huntingdon, David I owed allegiance to the king of England. He saw it as a personal relationship that did not include his kingdom. Traditional ties of loyalty and obligation were what defined people rather than national labels but that would change in the Middle Ages.

Before he succeeded his brother, David was known in surviving documents as the Prince of Cumbria, a territorial hangover from the faded kingdom of Strathclyde. And he had also taken control of the earldom of Northumberland, a vast domain reaching south to the Tees. England was divided by a civil war between King Stephen and the Empress Maud in the 1130s and David exploited this period of weakness to take parts of the north into his control. When Henry II became king in 1154, a year after David of Scotland's death, he quickly retrieved Northumberland and seized Cumbria. Scots kings such as William the Lion attempted to reassert themselves in England – in his case, with disastrous consequences as he was taken captive at Alnwick in 1174 and forced to recognise Henry II as his superior – but the southern borders of Scotland began to settle on the Tweed and along the Cheviot watershed.

In the west the macMalcolm kings were more successful. After initial forays into England and a brief period when the northern counties returned to his overlordship, Alexander II was forced to be more conciliatory towards England. From 1217, he turned his attention to the Atlantic west and quickly subdued Argyll. By the early 1230s, Ross and Caithness were brought more firmly into royal control and oversight, and with the death of Alan of Galloway in 1234 the Gall-Gaidheal began to become integrated into the kingdom.

Haakon's great warship had been built in Bergen entirely of oak and its prow was a gilded dragon's head. It would plough through the waves of the North Sea in the late summer of 1263 as the Norwegian king sailed with his war fleet to Scotland. After he had begun to rule in his own right in 1261, Alexander III had been pressing hard on Norwegian territory in the Western Isles. At last, his captains and strategists had understood the importance of sea power in the west and had begun to commandeer and build fleets of birlinns to attack and subdue the islands and coastal areas only easily accessible by ship.

Haakon was forced to act and, once he had forced Orcadian and Shetlandic shipmasters to join the war fleet, he sailed west and rounded Cape Wrath on 10 August. A rendezvous had been set with King Magnus of Man and the Norwegians arrived at a place whose name recalls the muster. Kyleakin on the eastern coast of Skye commands the entry to the natural harbour of Loch Alsh and its name is from *Caol Haakon*, 'the Straits of Haakon'.

The massed fleet of about 120 warships sailed south to the Firth of Clyde and anchored off the Cumbraes. A storm blew some ships onshore and Haakon's captains attempted a rescue near what is now the town of Largs. After a series of inconclusive skirmishes, Haakon retreated, turning his fleet north to overwinter in Orkney. When he unexpectedly died, impetus was lost and negotiations began for the Western Isles to be ceded to Scotland. Orkney and Shetland were to remain in Norwegian hands for two centuries more, only becoming Scottish in 1469.

These political shifts were significant. The Hebrides now looked south and east to Scotland and their old name, *Innse Gall*, 'the Islands of the Strangers', began to seem less fitting. Haakon had compelled the recruitment of Islesmen into his war fleet but most were inclined to look to the Scots king and authority passed to him – at least nominally. Trade and contact increased but, in the thirteenth century, most communities continued to live emphatically local lives as they had done for millennia.

On the night of 19 March 1286, there was a feast in the Great Hall of Edinburgh Castle. After carousing with his cronies, King Alexander III announced to the company that he did not wish to spend the night in his draughty castle on its windy rock. Instead he would ride to his manor at Kinghorn in Fife. There his new bride, the beautiful Yolande de Dreux, was waiting. Despite the lateness of the hour, a horse was saddled for the king and a party of escorts scrambled to make ready. By the time they clattered down the cobbles of Castlehill and the Lawnmarket, a storm was brewing and strong winds were blowing up the Firth of Forth.

When the king's party reached South Queensferry, the boatmen refused to cross. The wind was whipping spindrift off the waves and the sea was high. But gold glinted in the darkness and the oarsmen pushed off into the firth. By the time North Queensferry was safely reached and fresh horses found, the foul night had become a rainstorm. The burgesses begged the king not to go any further and they offered warm lodgings for the night. Probably still flushed with wine and desire and not used to being gainsaid, Alexander III insisted on pressing on through the black darkness.

He became detached from his party and, on the high ground above Pettycur Bay, only a mile or so from his manor and his bride, the king's horse lost its footing. As flight animals, horses lose their hearing in high wind and become frightened and easily spooked. As the wind and rain blew hard off the sea, a gust may have caught the animal broadside on. The king fell and could not be found on the path. Despite the efforts of search parties, his body was not discovered until morning. At the foot of the cliff, the last Gaelic king of Scotland, the last of the macMalcolm dynasty, lay dead. He had left no heir except a little girl, the Maid of Norway, and on her death the problems of the royal succession propelled

Scotland into a long and destructive period of war with England.

After the humiliating deposition of King John Baliol, the English began a series of invasions to bring the Scots more fully under their direct control. But these did not always go to plan. In June 1298, Edward I's army was starving and his men fighting with each other at their camp near Edinburgh. The Hammer of the Scots was about to order an ignominious retreat back to England when he was told that his enemies, led by the traitor William Wallace, had formed up for battle near the Wood of Callendar at Falkirk. Edward was exultant. 'As God lives,' he roared, 'they need not pursue me, for I will meet them this day!' He knew that his archers and armoured knights would win any pitched battle, and so it turned out. The Scottish schiltrons were thinned by a fatal rain of arrows before being scattered by the charging knights.

Amongst the Scottish dead was Sir John Stewart of Bonkyll, an estate in East Lothian. He had commanded the Scottish archers. It must have seemed like a terrible end, but in fact it was a beginning. Unknown to Sir John, a DNA marker had arisen in him that would genetically define Scotland's greatest royal dynasty. He carried R1b S781+, the marker of kings, the marker of the Stewart dynasty.

Scientists are certain about this because they sampled the DNA of descendants of Sir John's two sons and descendants of his brother, James, in the male line. The modern descendants of both of Sir John's sons carry the Y chromosome marker S781+ but the descendants of his brother, James, do not have it. By a straightforward process of deduction, this means that the marker arose in Sir John Stewart of Bonkyll and not in his father. If it had, the descendants of James would also carry it. And they do not. It is the first time it has been possible to link the genesis of a genetic marker to a named historical individual.

The second part of these findings is equally fascinating. Regardless of their family trees, 50 per cent of all men who have the surname of Stewart or Stuart are the direct descendants of Scotland's long-lasting royal dynasty (who also came to rule over Britain and Ireland). This data has been derived from sampling of the general population.

There are about 70,000 carriers of the surname in Britain, which means that about 17,500 men are of royal descent. Many of these live in Scotland, and most do not know that they have royal blood. And it is likely that some men who do not carry the Stewart surname will also carry the royal marker.

As further confirmation, scientists tested the DNA of Richard, 10th Duke of Buccleuch, looking at his whole genome. He is a direct descendant of Charles II (through his illegitimate son, James, Duke of Monmouth) and he did indeed carry the Royal Stewart marker that was passed down through the male descendants of Sir John Stewart of Bonkyll. But he is clearly not alone.

The reason there are so many descendants is very simple. In contrast to our (largely) monogamous society, powerful men had sex with many different women in the past, a phenomenon known as social selection. And for that reason their Y chromosome lineages spread very much wider than they do now.

Further research suggested that the Royal Stewart marker's historical journey was unexpected. Sir John Stewart's ancestor, Walter the Steward (it is an occupational surname), and his male ancestors came first to England with Henry I sometime around 1100. They had owned land in Normandy. There exists a plausible hypothesis that their ancestors had been part of the fifth- and sixth-century migrations from what is now the West Country of England. They fled as the Saxon invaders spread westwards. This possibility is founded on the modern distribution of the Stewart marker in that part of England. This prompted a national newspaper to make the appalling suggestion that Bonnie Prince Charlie was English.

The discovery of the Royal Stewart lineage did not, however, prevent a famous murder – although it did prove that it was motivated by a mistaken assumption. Henry Darnley's rashness was probably stoked into something worse by his conviction that the baby who would become James VI and I was not his. Perhaps the murdered David Riccio was the father. But ancestral DNA testing has proved that Darnley was indeed the father because he carried S781+. His suspicions were discovered to be ill-founded – 450 years later.

Recent sampling has shown that the Stewarts were not uniquely promiscuous. It appears that more than 10 per cent of all Scottish men are descended from patriarchs of some kind. A few are known, but many are not. The Gaelic word *clann* means children and it seems that several clans are indeed the children of a single individual and these groups can be substantial. The R1b-S690 Y chromosome marker is carried by many MacGregors, including their chief, Sir Malcolm. But what is striking is how many Scotsmen have it – more than 25,000. The patriarch who founded this old lineage may have been Iain Cam MacGregor and he lived in the second half of the fourteenth century. Many MacFarlanes, Hamiltons, Frasers and MacLeods are also descended from a single individual. But some of these lineages might disguise a very intriguing link.

Seven clans call themselves Siol Alpin, the Seed or Descendants of Kenneth MacAlpin, the famous early king of Scotland – the first, it is still claimed, to unite the kingdoms of the Picts and Scots, and the king from whom all Scottish monarchs are numbered. The link to Siol Alpin has long appeared to be more a tradition than a fact, but recent DNA testing shows that at least five are indeed genetically linked: MacGregors (whose motto is 'S Rioghal Mo Dhream, My Race is Royal), Clan Grant, MacAulay, MacFie and MacKinnon. The remaining two, MacQuarries and MacNabs, do not appear to be linked, but more testing is probably needed. If these and other links are confirmed as more samples come in, then it may well be that many Scotsmen are indeed of a royal race.

When Robert the Bruce emerged as king, a Norman-French family had replaced a native dynasty. Royal DNA had changed but that of the people did not. The Bruces had been in Scotland for two centuries, and on their accession there was no great in-migration of new people, only a powerful assertion of independence from England – and a fascinating linguistic and cultural shift.

Roxburgh is a strange, atmospheric place – a lost city of medieval Scotland. It gave its name to an old county, a ducal title and to a small village two miles to the west of the original site. Founded on a river peninsula formed where the Teviot joins the Tweed and lying below the ancient fortress of Hroc's Burh, it was founded by David I 'when he was an earl' some time before 1113. It became the prime commercial hub of Scotland. And yet there is now no trace of it, not a single stone left standing on another. Well recorded in the documents left by the Border abbeys, Roxburgh now exists only in ink and on vellum, its streets and markets described in monastic Latin. War destroyed the city, sweeping away its houses amidst fire and sword. During the years of war between England and Scotland, it lay too near the border to be allowed to thrive.

In the early twelfth century, the significance of the new burgh was immense. Representing the first genuine stirrings of urban life in Scotland, it grew quickly as the wool trade developed. Part of David I's motivation in founding and endowing the Border abbeys so generously was to exploit one of the most productive areas of his kingdom properly. The monks ran huge and efficient sheep ranches in the Cheviots and Lammermuirs and merchants from England, Flanders and Italy came to Roxburgh's markets to buy their wool crop.

The new town lay firmly in the heart of old Bernicia and the farmers and shepherds who produced the wool had spoken English for many generations. Trade quickly generates comprehension and the language of the producers and sellers became a lingua franca. Money talked English. Coins were first minted in Scotland at Roxburgh and, with a large royal castle immediately to the west and the richest abbey in the country to the east at Kelso, the burgh became the lynchpin of David I's vibrant new economy. The only place of comparable size and importance was the busy port of Berwick at the mouth of the Tweed. It had also been Bernician and had also spoken English for many generations. Between the two towns ran the Via Regis, a royal road down which trade flowed. And so did language, the English language.

At the court of Alexander III, it is likely that Gaelic was spoken, but by the fourteenth century it had been supplanted by English – or at least a Scots version of English. King Robert the Bruce had lands in Celtic Galloway and was probably fluent in both but Scots English had become the language of business – and also of poetry.

John Barbour was a priest at Dunkeld Cathedral in 1356 before being promoted to the archdeaconry of Aberdeen. He spoke and, crucially, wrote in Scots. While serving at the royal court of Robert II, Barbour composed a long narrative poem to celebrate the achievement of the dynasty. *The Brus* begins with the death of Alexander III, makes the Battle of Bannockburn its centrepiece and concludes with the burial of Bruce's heart at Melrose Abbey. 'A! fredome is a noble thing' is perhaps the most famous line of this patriotic epic.

The likely author of several other long poems in Scots, Barbour appears to have been tremendously prolific – *Legends of the Saints* runs to a staggering 33,000 lines. Also significant were Barbour's origins. He was attached to the royal court but appears to have had no connections with the Lothians or the Borders, the early heartlands of the Scots language. David I's early burghs had been added to in number all over the map of Scotland and the efforts of succeeding kings to tame the north had involved plantations of communities from the south. Just to the east of Nairn, at Auldearn, there exists a linguistic frontier that is still audible. To the west the Inverness accent derives from people who were native Gaelic speakers and who learned English in the modern period. The distinctive Doric of Scots spoken east of Auldearn and along the Moray coastlands down into Aberdeenshire was originally the speech of communities that were transplanted north from the areas where the language was commonly spoken in medieval Scotland.

Language and DNA often move together but, as time wore on, the pattern across Scotland grew more settled. People began to turn to other indicators when they considered their ancestry and appearance is still cited as one of the most obvious and important. Across the diversity of African DNA, it is manifestly true that there exists a wide range of physical types, from the tall tribesmen of Kenya and Nigeria to the pygmies and the bushmen of the Kalahari. But, in the much more restricted scope of non-African DNA, variables are less obvious over fairly large areas such as the European peninsula.

Clearly the Chinese look different from Indians who do not look like Europeans and they, in turn, do not resemble Native Americans very closely. Skin colour and eye shape are the most dramatic differences across the world outside of Africa, but within Europe the diversity between north and south is sometimes difficult to see clearly, especially if a suntanned skin is discounted. Within Scotland the classic indicators are popularly seen as hair and eye colour, with some modifications. But, in reality, most Scots look as peely-wally as each other and skin colour varies more with the seasons than DNA, from scarlet to porcelain-white. Clearly, this uniformity has been altered by recent immigration from further afield but, for the moment, that story lies in the future.

Tacitus described the red hair and large limbs of the inhabitants of Caledonia, believing that they looked Germanic. When Dio Cassius wrote of the rebellion of Boudicca in AD 60, he waxed operatic and she was apparently 'tall and terrifying in appearance . . . [with] a great mass of red hair . . . over her shoulders'. These observations are obviously anecdotal and Tacitus's views were coloured by his own experience of Germans and Germany. But those references to red hair are, in fact, very apposite in the assessing the modern picture.

Per capita, in every sense, Scotland has the most redheads in the world. Across the entire human population, only 1–2 per cent has red hair in its various shades. But in Scotland 13 per cent of our five million are redheads and approximately 30 per cent carry one of the variants needed to pass on that characteristic from one generation to the next. While redheads carry two of these variants, people who carry just one can have hair of any colour. But they can pass that variant on to the next generation and, if a couple both carry one variant, then, on average, 25 per cent of their children will have red hair.

Most red hair in Scotland is caused by a combination of three fairly common variants in the same gene. The gene encodes a hormone receptor which controls whether black/brown or red pigment is produced. The variants are very ancient, at least tens of thousands of years old and part of our shared European hunter-gatherer heritage. The intriguing question is why red hair has become so common in Scotland. It has been suggested that it is linked to climate. More surprising is recent research showing that some Neanderthals had red hair because of different variants in the same gene. What all that adds up to is the remarkable fact that 650,000 Scottish men and women have shades of red, auburn, strawberry blond and all of the glorious tints between. Closest statistically are our Celtic cousins. Approximately 420,000 Irish men and women and 290,000 Welshmen and women are similarly blessed. In England, the percentage is thought to be lower at around 6 per cent but that means a total of three million redheads in a total United Kingdom and Ireland population of 4.6 million, a substantial disincentive to those foolish enough to mock – an army of redheads large enough to impress Tacitus.

Redhead carrier frequencies across Britain and Ireland (based on the three most common redhead variants)

More recent research has supplied a clear sense of the distribution of the red hair gene variant carried by many Scots (who do not necessarily have red hair) and its nature. Red hair is passed on in DNA through three common variants. Using the names of the biochemicals that characterise each one, the provisional proportions are:

Cysteine-red (or R151C) is carried by 10 per cent of British people

Tryptophan-red (or R160W) is carried by 9 per cent of British people

Histidine-red (or D294H) is carried by 2.5 per cent of British people

There are about 40 other, much rarer variants. Everyone who carries one of these variants is a direct descendant of the first person ever to have it. Those with Cysteine-red have a 70,000-year-old variant that probably arose in West Asia, those with Tryptophan-red are the descendants of someone who probably also lived in West Asia 70,000 years ago, and finally Histidine-reds belong to a much younger group who descend from a European who lived about 30,000 years ago.

A very large sample of 5,000 individuals yielded much more precise information about the distribution of the red hair gene variants across Scotland, the rest of Britain and Ireland. It is very revealing and runs counter to many assumptions. Across Britain and Ireland the average distribution amongst all people is 33 per cent, likely to be the highest proportion of any population in the world. How the nations of Britain and Ireland break down is also surprising. The average distribution of the red hair gene variant in England is the lowest at 28.5 per cent, with the east of England the lowest region in Britain and Ireland with only 21 per cent carrying it. At 38 per cent, Wales is considerably higher, as is Ireland at 36.25 per cent. The average for Scotland is the lowest for the so-called Celtic nations at 35.4 per cent, but the variations within the country are wide. In the Northern Isles, the Hebrides and the Atlantic coast, only 29 per cent carry the variant, whereas the highest regional percentage in Britain is in Edinburgh and south-east of Scotland at 40 per cent of all people.

These statistics add credibility to the climate-based hypothesis that explains the distribution of redheads and the red hair gene variant. We all need Vitamin D from sunshine and that it why most Africans are dark-skinned and most Europeans are lighter skinned. It is a question of balance: too much sun in Africa, too little in much of Europe. Red hair and the carriers of the gene variant should, therefore, occur more frequently further north in Europe. Not exactly. That would mean most redheads in Scandinavia, but that is not the case. Most are in Britain. The simplest explanation might be the best. Where there are most redheads, according to present statistics, in Scotland and the north of England, there is much more cloud each year than sunshine. In Sweden, for example, the average daily number of sunshine hours is 5.4, while in Scotland it is 3.1. There is an argument that redheads are well adapted to their circumstances, perhaps better adapted than those with other hair colours. This might be seen as a good rejoinder to those who talk of 'gingers' in a disparaging manner.

The link with Neanderthals has also now been discounted. Archaeologists have discovered about 400 Neanderthal skeletons across Europe and the Near East. DNA testing shows that some had red hair and fair skin. But the variant was different from that seen in modern humans, and although we cross-bred with Neanderthals, early speculation that they were the source of the red hair in modern Europeans is now thought to be just that.

Tacitus and Dio Cassius were not only the first to note red hair in Britain, they were also amongst the originators of making a link between it and a fiery disposition. This is, of course, unscientific and ahistorical but it is a persistent and corrosive myth, part of a Celtic cliché. One rugby commentator once shamelessly described a red-haired Irish rugby player as temperamental – 50 per cent temper and 50 per cent mental.

An additional trope is related to the belief that Celts are small and dark. One of the earliest to examine this and other issues of ethnicity in Britain and Ireland was a Victorian scientist, John Beddoe. After graduating in medicine from Edinburgh University, he set up in practice in Bristol. There he began to work on compiling scientific data on the appearance of the population of both countries (united at that time). Concentrating on hair and eye colour, he devised a formula called the Index of Nigrescence which accounted for a range of hair colour from jet black to fair and red. Travelling all over Britain and four times to Ireland (even sailing out to the windswept Aran Islands in a heavy sea), Beddoe built up a mass of data. In 1885, he published *The Races of Britain* and it was well received by both the public and the academic world – a rare thing.

There are many maps in Beddoe's book and, over a lifetime, he had collected data on a huge sample of 43,000 individuals. What his Index of Nigrescence showed across the map of Britain is fascinating and, leaving aside the outdated Victorian notions of ethnicity that accompanied the work, it is still valuable. On hair colour alone, there is a wide divergence from east to west. In Kent, East Anglia and for most of the length of the North Sea coastline, Beddoe found that there were as many people with fair and red hair as there were with dark hair. The same picture emerged in the Northern Isles and in the Outer Hebrides. But as he moved west-wards, Beddoe found that dark hair outnumbered fair and red in Cornwall, Devon, Somerset, Dorset and Wiltshire, Wales, Galloway and Argyll. In Ireland, dark hair was generally very much more prevalent with the greatest concentration in the west.

Properly shy of making hard and fast conclusions from all that work and the many miles he had tramped, Beddoe did allow himself some speculation. He thought that people in the east of Britain derived their ancestry from the Anglo-Saxons and the Scandinavians and that, in the greater part of England, it amounts to something like half. In Ireland, the Atlantic west and Wales, he believed that there had been an in-migration from the Iberian Peninsula. Although the modern DNA picture introduced a great deal more complication and sophistication, when it is laid over Beddoe's work, there is more than a passing correlation. Hair colour turns out to be more than folk genetics.

Eye colour has also been the subject of rigorous recent research. For most of prehistory, there was no variation: amongst *Homo sapiens*, everyone had brown eyes. And most of the world's popula-tion still does. Around 10,000 years ago, almost certainly around the shores of the Baltic Sea, a variant in the eye colour of a single individual appeared. He or she was the first to have blue eyes and Pierce Brosnan, Kate Winslet and all those who have baby blues are descended from that person.

In a matter of only a few thousand years, the most common eye colour frequency in northern Europe flipped from brown to blue.

And a unique map shows the distribution of blue eyes in Europe. The key shows a staggering 80 per cent around the Baltic, with the highest at 99 per cent in Estonia. And within Britain, in Scotland, Ireland and the regions of the north of England, there are from 49 per cent to 79 per cent with blue eyes, and in southern England and Wales significantly fewer, with 20 per cent to 49 per cent.

The genetics of eye colour are complicated. Unlike red hair (which is a recessive trait), eye colour is polygenic: it is determined by the interaction of several genes.

This variant alters the production of the pigment known as melanin. Melanin in the iris affects how much light is absorbed and how much light is reflected – the more melanin that is present, the more light is absorbed. People with blue eyes actually have a very low amount of melanin in their irises, technically a lack of colour. When light hits the eyes of blue-eyed people, the longer wavelengths pass through and are absorbed in the back of the iris, while shorter wavelengths scatter and reflect back out into the open air. Since the shorter wavelengths are blue, people will view the iris colour as blue. When there is a high amount of melanin in the iris, both the long and short wavelengths of light are absorbed, so there is no reflection of light out of the eye; this results in brown eyes. When the amount of melanin is somewhere in the middle, between blue and brown, a green eye colour is seen.

So, how does one variant affect melanin pigment? The gene in which the mutation occurs, HERC2, acts as a regulator for the melanin production gene, OCA2. When the ancestral or brown form carried by everyone until 10,000 years ago (AA) is present, OCA2 is at full production and enough melanin is produced to cause brown eyes. But when the variant form is present and A is replaced with G, HERC2 acts as an 'off-switch' and melanin production becomes severely limited. This 'switch' is particularly effective, as it still leaves a small amount of melanin so the eye retains functionality. If melanin production was completely shut down, rather than limited, it would result in albinism, where people popularly known as albinos have pink eyes.

There are many theories as to why the G or blue variant became so common across Europe. The most likely is that an eye colour that stood out from the crowd enabled individuals to secure mates more easily. This so-called sexual selection typically favours bright colour traits, particularly those that are different from the norm – examples include the extravagant plumage of the birds of paradise or the manes of lions. The increased presence of blue eyes rose too sharply for chance effects (known as genetic drift) to be a cause, and there is also no survival advantage to having blue eyes. You cannot see better than people with brown or green eyes. Perhaps, just like the peacock's tail, it persists because it is attractive to mates. And because people with blue eyes are more successful at attracting mates, a snowball effect ensues.

Very many successful actors and actresses have blue eyes and the reason why Cameron Diaz, Brad Pitt, Cate Blanchett, Angelina Jolie, Nicole Kidman, Hugh Grant, Kate Winslet, Keifer Sutherland, Charlize Theron, Matt Damon, Jennifer Aniston and Pierce Brosnan all have blue eyes may be linked to the fact that they appear to sparkle. This is the effect of the reflection of shorter light wavelengths.

An analysis of the genetics of eye colour was recently carried out and it looked at how it behaves in a sample of several thousand British men and women. Here are some fascinating findings.

- Ireland has the highest percentage of people with blue eyes.
- In Britain and Ireland, the region with the highest percentage with baby blues is Edinburgh and south-east Scotland.
- In Britain and Ireland as a whole 48 per cent have blue eyes, 30 per cent have green, and 22 per cent brown.

Folk memory and myth-history can be a great deal less reliable. A frequently heard explanation for so-called Spanish or Iberian looks in the west of Scotland, Ireland and elsewhere is the entertaining notion that many dark-haired people are descended from survivors of the shipwrecks of the Spanish Armada of 1588. This widely held notion does not survive the briefest test of statistics and common sense.

One hundred and forty-one ships set sail under the command of the Duke of Medina Sidonia with the intention of facilitating a landing in England by soldiers from the Spanish possessions in Flanders. After being engaged by the English fleet in the Channel and losing several ships, the Armada was chased north and forced to circumnavigate Britain. In the teeth of fierce Atlantic storms, more than 30 ships were wrecked, run aground or forced to seek shelter. Of these, 24 certainly foundered off the Irish coast and perhaps only six or eight off the coasts of Scotland. Most sailors drowned or were killed or imprisoned by local populations. Any Armada survivors are likely to have saved themselves after shipwrecks or scuttlings and only a tiny number, perhaps a dozen – if any – will have somehow made it to shore. And it must have been a shore away from habitation for them to avoid capture and possibly death. As these wretched sailors scrambled over rocks and beaches, soaked, exhausted and probably injured, unable to utter a word of Gaelic, sex with a local lass would have been the very last thing on their minds.

Such a sequence of events is not fanciful. For large numbers of people in western Britain and Ireland with dark hair and so-called Iberian features to claim Armada survivors as ancestors – that is something close to what will have had to happen. And, within a large local population, a tiny number of men will need to have fathered a prodigious brood of children carrying their DNA. Very unlikely indeed. And no genetic evidence has been found to support Armada ancestry in any Scottish – or Irish – person tested.

Wha's Like Us?

✼

I N THE SUMMER OF 1315, flushed with victory at Bannockburn the year before, King Robert the Bruce led a Scots army across Hadrian's Wall to attack the walled city of Carlisle. Its defences were marshalled by Andrew de Harcla, a fascinating figure, and, when scouts warned him that the Scots were close, he ordered the bridge over the River Eden to be torn down. When Bruce's forces reached Stanwix Bank and the old Roman cavalry fort, they saw an awkward obstacle between them and the city. The Eden was running in spate, a roiling, brown torrent of floodwater. But Bruce's engineers managed to rig up a wooden bridge and eventually they rumbled across.

De Harcla's garrison of 500 or so, mainly archers, was massively outnumbered and the very long walls of Carlisle were difficult to defend against simultaneous assaults in several places. But that did not matter. Throughout an 11-day siege, the Scots were never able to reach the walls. The weather defended Carlisle. Here is an extract from the *Lanercost Chronicle*:

> Moreover the Scots had many long ladders, which they brought with them for scaling the walls in differently places at the same time; also a sow for mining the town wall, had they been able: but neither sow nor ladders availed them anything. Also they made great

numbers of bundles of corn and herbage to fill the moat outside
the wall on the east side, so they might pass over dry-shod. Also
they made long bridges of logs running on wheels, such as being
strongly and swiftly drawn with rope might reach across the width
of the moat. But during all the time the Scots were on the ground
neither bundles sufficed to fill the moat, nor those wooden bridges
to cross the ditch, but sank to the depths by their own weight.

This was one of many reports of the beginning of another Ice
Age, which historians have come to call the Little Ice Age. For a
long time, in more or less extreme episodes between 1300 and 1850
there was a substantial deterioration in the weather, characterised by
cold winters and wet summers. When Bruce's men were flounder-
ing in the mud at the foot of Carlisle's walls, the corn was rotting
in the fields, soaked and green, deprived of the ripening sunshine.
The harvest failed again in 1316 and famine spread like a cancer,
killing first the old and sick and then the children.

Conditions grew steadily worse in the fourteenth century and
the Norse colonists were forced to abandon their settlements in
Greenland, those they had established in the sunlit centuries before
1300. Measurements taken from ice cores pulled out of Alpine glaciers show significant surges of bad summers and very cold winters
in the 1590s, 1690s and 1810s.

Pieter Bruegel's famous painting *The Hunters in the Snow* shows a
memorable winter landscape, a scene familiar to all who looked at it
and shivered. In the foreground, a band of hunters trudge through
deep snow with their dogs. They look out on a frozen lake where
skaters skim across the surface. Beside a river is a watermill, its
wheel stiff with frost.

When William Shakespeare's plays were first performed at
London's Globe Theatre on the south bank of the Thames, it was
possible to walk across the frozen river to see them. So intense was
the cold in the winters around 1600 that frost fairs were held on
the Thames even though the tidal river has a high salt content as it
flows through London.

Poor harvests and dismal weather no doubt made a hard life for
Scotland's ordinary people even harsher. But much worse was to

come. By June 1348, the Black Death had reached the southern coasts of England, taking its first victims at the village of Melcombe in Weymouth Bay. Fleas from black rats and humans carried the deadly bacillus and when they bit they regurgitated it directly into the bloodstream. With swollen lymph glands under the armpits and in the groin, those infected died in agony – but mercifully quickly, usually after only four days.

By 1349, the Black Death had yet to reach Scotland and the less sensible believed that it did not affect Scots, presumably because . . . they were Scots. In order to take cruel advantage of the stricken English, an army mustered at Caddonlee near Galashiels. But around the campfire several men suddenly became ill, the plague erupted in the midst of the muster, probably brought by mercenary soldiers, and the army panicked and fled.

After that initial outbreak, the disease spread like wildfire as soldiers returned home. At least a third of the population of Scotland died, the densely settled agricultural districts as badly affected as the towns. Only in the isolated glens of the Highlands and in the islands did communities escape – at least for a time. The devastation of 1349 was the most severe visitation but not the last. The Black Death stalked Scotland again in 1361, 1379, 1392, 1401–1403, 1439 and 1455. It is highly likely that Scotland's collective DNA was altered by such an extreme death rate and several lineages may have died out. The impact of such carnage, unlike anything sustained in war, coupled with sustained climate change was, of course, catastrophic. As agricultural output declined so did trade and a long recession descended on Western Europe.

In Scotland, power politics made matters worse. After the death of Robert the Bruce, a long civil war festered as the heirs of John Baliol pressed their claims to the throne with the backing of meddling English kings. As a succession of weaker monarchs attempted to assert authority, local lordships waxed stronger. The Douglases, the Kers and others compiled large fiefdoms and regularly ignored or challenged kings. With the beginnings of the Stewart dynasty in the ineffectual shapes of Robert II, his son Robert III and the governorship of the Duke of Albany until 1420 (while James I was a minor and uncrowned), Scotland seemed to stand still, perhaps even regress. On his death in 1406, the miserable Robert III left an astonishing, self-pitying epitaph: 'Here lies the worst of kings and the most wretched of men in the whole realm.' The impression is of a backward, colourless corner of north-western Europe.

As the fifteenth century wore on the population of Scotland began to recover, although it took a long time to reach the early medieval optimum of around a million souls. And the Stewart kings embarked on a long war with some of their own subjects. Highlanders were seen as very different – primitive, savage speakers of another language, often called 'Irish'. Walter Bower related a remarkable incident that took place on the North Inch, an island in the Tay at Perth. Two Highland clans had been unable to resolve a dispute amicably and it was agreed that the affair should be decided by judicial combat in the presence of the king. Grandstands were built, English and French spectators flocked to Perth and the events of the day were widely reported. Thirty armed men from each side came to the North Inch and none were allowed armour. This was to be a fight to the death:

> At once arrows flew on either side, men swung their axes, brandished their swords and struggled with each other; like butchers killing cattle in a slaughter-house, they massacred each other fearlessly; there was not even one amongst so many who, whether from frenzy or fear, or by turning aide from a chance to attack another in the back, sought to excuse himself from all this slaughter . . . and from then on for a long time the north remained quiet, and there was neither evil nor upset there as before.

There is an unmistakable sense here of Highlanders as lesser beings – like animals, in fact, as only 12 men were left standing amongst the terrible carnage and the groans of 48 dying men. The horrors of the Colosseum in Rome come to mind.

The Stewart kings were ruthless in their 'daunting' of the clans and none more so than James VI. His Statutes of Iona in 1609 forced chiefs to have their eldest sons and daughters educated on the mainland and it attacked Gaelic language, manners, dress and customs. Nine chiefs had been tricked, abducted and compelled to sign the document. It was the beginning of a shift in power from the Highlands and, eventually, a shift in population.

In the opening decade of the seventeenth century the British crown also undertook a concerted police action at the other end of Scotland. The Border Reivers were also to be daunted and a series of mass arrests and executions took place. It has long been asserted that the reiving surnames were not like the Highland clans and that those Borderers who have formed clan associations in recent times misunderstand their heritage. But DNA research may prove them right after all.

New genetic evidence shows that the Armstrongs on either side of the border are much more like a Highland clan than a mere surname, and that they did not originate in the Borders. The Gaelic term *clann* is again useful here, as it turns out that the Border Armstrongs are indeed children – the children of one man.

In essence, a new analysis has discovered that the haplogroup the Armstrongs descend from in the male line originated not in the Borders but in ancient Pictland. Their distinctive Y chromosome DNA of R1b –S389 rises to above 1 per cent of all men in Perthshire, Stirlingshire, Tayside and the north-east of Scotland, and it arose perhaps 3,000 years ago. At some point in the historic period, perhaps around a thousand years ago, a single individual, one man, moved south to the Borders, almost certainly to Liddesdale. There he became powerful, probably acquiring land or property, and he also became the progenitor of the great reiving family. How is it possible to be sure about this?

Geneticists can recognise what are known as expansion clusters, groups of men who carry closely related DNA markers that point to a common male ancestor. And just such a cluster was recognised in men who carry the Armstrong surname and who originate in the Borders. What makes these groups stand out is the fact that there are many more of them than might be produced by the normal genetic inheritance process. Men have sons, or not; sometimes several, sometimes one. Surnames stay alive and the numbers who inherit them grow slowly. But these expansion clusters appeared to increase quickly, and they are also often young, probably about 1,000 to 1,500 years old. Their recent appearance and the rapid increase in numbers all point to the operation of what is known as social selection, or breeding advantages.

That is a polite term for the fact that kings, princes and other powerful men in the past had sex with several, often many, different women. The original Armstrongs must have been able to access several women – in Liddesdale or wherever they first settled. In comparison to our largely monogamous society, these men were able to spread their ancestral DNA very wide and have many sons. They in turn often became powerful and repeated the process.

This is the pattern of genetic inheritance that happened with the man who came from ancient Pictland to settle in the Borders – the first Armstrong, and the begetter of not only a great name but also men who shaped Scottish history. The Highland clans were feared because they formed a family army and fought ferociously for each other, men of the same blood. The Armstrong Heidsman could put 4,000 men in the saddle in a morning and wherever they rode, to fight or raid, they too were feared. Perhaps for the same reason. They fought not only for their name, but also for their blood.

Fifty years before the Iona Statutes were enacted, an even more far-reaching revolution convulsed Scotland. After 1560, the Reformation had succeeded and Catholicism was in widespread retreat. The abbeys and convents that had first developed and then owned much of the Scottish countryside were taken over, many by secular lords who set about building up large estates. The patrimony of the Dukes of Roxburghe, for example, is largely the estate owned by the wealthy abbey of Kelso.

The reforms of John Knox and others began to acquire momentum and, by the end of the sixteenth century, clergymen were proclaiming the Scots as a chosen people who lived in Christ's Kingdom of Scotland. At Falkland Palace, Andrew Melville famously reminded James VI in 1596 that:

> [t]hair is twa Kings and twa Kingdomes in Scotland. Thair is Christ Jesus the King, and His kingdom, the Kirk, whase subject King James the Saxt is, and of whase kingdome nocht a king, not a lord, not a heid, but a member.

As well as a tendency to hector, a central tenet of the reformed religion was the priesthood of all believers and this involved a huge national commitment to literacy. The members of Christ's 'Kingdome in Scotland' had to be able to read the Bible, the Word of God, for themselves without the need for the 'mumbojumbo' of a priest. Schools were eventually built in every parish in the land and, in order to train enough ministers, there were five universities (two in Aberdeen) while England had only two. For once, politics reached down to ordinary people and, with the advent of mass literacy, changed their lives for the better. The traditional Scottish reverence for education and literature is a lasting legacy.

By the time schools were well established in the seventeenth century and the five universities were producing graduates, a surplus of highly educated Scots emerged. A brain drain began. Many left Scotland to find work in England; some sailed east to the Baltic and, of those, significant numbers settled in Poland. It is thought that by 1800 more than 30,000 lived there, mainly in the cities of Gdansk, Kraków and Warsaw. They formed Scottish Brotherhoods and maintained links with their homeland but most never came back. In the twentieth and twenty-first centuries, Poland was to return the compliment.

As a separate nation under a united monarchy, Scotland had, in some ways, the worst of both worlds. England developed its colonies in North America with growing success but Scots had very limited access to all that potential. After the convulsions of the Civil War, the invasion of Cromwell and the bloodshed of the Wars of the Covenant, some Scots were much moved to seek a life elsewhere. Many went out to Virginia, but not as colonists. Most were indentured servants, virtual slaves obliged to work for only food and shelter for a set period, usually seven years. On their release from these punitive contracts, some Scots succeeded in the New World. In 1684 a former messenger from Selkirk, Peter Wilson, wrote home to his cousin, 'Poor men like myself live better here than in Scotland, if they will but work.' He rented 25 to 30 acres at an annual rent of around five shillings and had a share of the crops.

Access to England's colonies became more problematic at the close of the seventeenth century as Scotland's parliament maintained her independence. In the 1690s, Scots attempted to take matters into their own hands. A 'Company of Scotland tradeing to Affrica and the Indies' was formed and a plan made to found the first colony of the Scottish Empire on the Isthmus of Panama, at a place called Darien. It was hoped that the venture would re-animate the moribund economy and it attracted widespread backing. Almost a quarter of the nation's liquid assets was poured into the Darien Scheme, and as ships cast off from Leith, spirits were high.

Darien was an immediate disaster. New Caledonia turned out to be a coastline of mosquito-ridden swampland peopled by hostile natives. Three quarters of the colonists died, the Company of Scotland was bankrupted and the nation plunged into despair and economic crisis. A series of poor harvests in the rain-soaked 1690s added to the atmosphere of gloom.

With the expulsion of James VII in 1688, King William of Orange and Queen Mary had acceded to the throne of Great Britain and Ireland. A year later there was a serious rebellion in the Highlands led by Viscount Dundee, a Stuart supporter. He was a charismatic general but was killed at the moment of victory at Killiecrankie. Jacobitism was seen as an ever-present threat and fear of it led King William into criminal excess. The royal signature appeared on orders that led to the commission of an atrocity in the north. In 1692, Captain Robert Campbell of Glenlyon, acting on secret orders, massacred 38 members of the MacDonald clan of Glencoe. Women and children, stripped naked by the government soldiers, it was said, later died of exposure in the snows as the houses of their settlements blazed in the glen below.

The incident sparked outrage and fuelled Jacobite sympathies. Highlanders could be slaughtered like cattle and attitudes appeared to have changed little since the carnage on the North Inch at Perth. The Glencoe atrocity added to a momentum gathering behind political as well as dynastic union with England. The economic problems heightened by the Darien fiasco also propelled Scotland to seek a brighter future as part of the state of Great Britain as well as the kingdom. Despite popular objection, especially on the streets of Edinburgh, the Scottish Parliament voted itself out of existence in 1707.

Those who plotted and were bribed to ensure the passage of the

Act of Union hoped that the Scottish economy would prosper in the wake of England's imperial ambition. It took time for output and trade to revive as they found markets in the south, but one of Scotland's most significant exports was her people.

Emigration often took place in stages. The eighteenth century is usually characterised as an age of agricultural improvement.
This certainly involved the invention of new and better machinery, most notably the revolutionary swing plough first used in Berwickshire, but it also meant what would now be called rationalisation. Land ownership and land use both changed.

Many of Scotland's small towns and villages enjoyed the use of common land and indwellers had ancient rights to pasture domestic cows, sheep and goats on these large communal holdings. They also cut peats, brackens and other useful items. But, as land values rose and utility was improved, private landowners often claimed and sometimes simply seized common land. As Baron of Hawick, the Duke of Buccleuch asserted that he owned the entire common. Legal arguments were wearily exchanged over a long period and eventually he was awarded a third of what the people of Hawick held in common. The remainder was enclosed with fences and dykes and some of it rented to individuals.

Not far away stands a poignant memorial to the enclosure of Scotland's arable land and its passage into largely private hands. In the seventeenth century, the prosperous Roxburghshire village of Longnewton had applied for burgh status. Now all that remains is a burial ground. Surrounded by a drystane dyke and sheltered by a stand of ancient trees, a few tumbledown headstones can be found in the long grass.

Longnewton disappeared in the eighteenth century when the Earl of Lothian decided to combine a number of smallholdings around the village into a new and much larger farm. Production would be much improved and value added if agriculture was conducted not for subsistence but on a more expanded, more practical scale. John Younger was born in Longnewton and he later recalled what happened. His father had 14 acres when his tenancy was terminated by the Lothian estate but they allowed him to stay in his house because he was a shoemaker, a useful trade. The dykes of the smallholding were thrown down and the land ploughed right up to the walls of the shoemaker's house. He had no garden and even his hens were shot.

Twenty families lived at Longnewton and, in a very short time, they dispersed. All over eighteenth-century Scotland similarly convulsive changes were being forced through as country people began to migrate to larger towns and cities in search of work. These early clearances took place in the Lowlands, beginning almost a century before the Highlands was seriously depopulated. What made this population shift less dramatic and less clearly remembered in history books was the fact that it seemed less drastic. Most people did not, like the former messenger from Selkirk, Peter Wilson, leave Scotland but were able to maintain family connections and community links, albeit over a distance.

The *First Statistical Account* was compiled by parish ministers in the 1790s and it mourned the passing of smallholders, people they called cottagers. At Kilmany in Fife, the account did record what seemed a profound change – 'the annihilation of the little cottagers' – and in Angus the effect was seen as even more emphatic – 'many of the cottagers are exterminated'. In a less overstated reality, there was employment available and few will have been rendered destitute. The new agricultural methods, particularly the swing plough and the management of the heavy horses needed to pull it, called for year-round contracts or fees, as they were known. And, further afield, there was a great deal of construction work beginning all over Scotland as the country began to urbanise. The most dramatic cumulative impact of the clearances across Scotland was seen in Glasgow. From the end of the eighteenth century until 1825, the city's population exploded, rising from 70,000 to 170,000 in a generation. The hardy sons of the soil who had grown up on the smallholdings disappeared into the streets of Scotland's large towns and cities as the old life on the land passed into memory.

The enclosure movement had driven up the cost of rural rentals for many larger-scale farmers, in some places by as much as 400 per cent. To many, emigration seemed a better option than labouring work digging ditches or making roads or moving in to the growing towns and cities. Friendly Societies were founded and by 1750 they were enabling emigration in numbers from Galloway,

Ayrshire, Stirlingshire and Perthshire in particular. Undertakings were signed and promises made. The societies employed agents in North America to find good land and negotiate for it. Once a deal had been struck, farmers boarded ships with their families and set sail for a new life. An advertisement in *The Gentleman's Magazine* of 1749 painted an attractive picture:

> Let's away to New Scotland where plenty sits Queen
> O'er as happy a country as was ever seen
> And blessed her subjects, both little and great
> With each a good house and a pretty estate.
> No landlords are there the poor tenant to tease
> No lawyers to bully, no Bailiff to seize.

But no pretty estate awaited a group of 300 emigrants from Killin near Stirling. On landing in Nova Scotia in 1776, they discovered that their fertile farmland was uncleared forest. If parties of emigrants arrived in the autumn or winter, too late to plant, they often had to depend on the kindness of their neighbours, if they had any, for food and shelter.

In the aftermath of the Jacobite rebellions of 1715 and 1745, Highlanders could scarcely expect much sympathy from their neighbours in Lowland Scotland. After the defeat at Culloden, the government army itself evicted supporters of Prince Charles and ministers at Westminster attempted to defuse any further trouble in the north. A blood price for rebellion was extracted when clansmen were recruited into the British Army. Between 1756 and 1815 at least 40,000 and perhaps as many as 75,000 clansmen marched to war in Europe, America and India in the great drive for empire. William Pitt, the Prime Minister, took a cynical view of the new recruits: 'The Highlanders are hardy, intrepid, accustomed to rough country, and it is no great mischief if they fall. How can you better employ a secret enemy than by making his end conducive to the common good?'

As their young men departed for foreign fields, the clansmen and -women left at home were serially betrayed by many of their chiefs. A bard of Clan Chisholm made an unequivocal point: 'Our chief has lost his feeling of kinship, he prefers sheep in the glen and his young men in the Highland regiments.' Not the most brutal but perhaps one of the most cynical was the grandly named Colonel Alasdair Ranaldson MacDonell of Glengarry, a MacDonald chief. In 1812, Henry Raeburn painted him resplendent in his plaid, kilt and feathered bonnet, holding a musket and staring into the middle distance. There, he might have seen his bailiffs evicting his tenants, families who bore his name and had lived in Glengarry since time out of mind. Despite forming The Society of True Highlanders, MacDonell was anything but, felling the ancient oak forests of his patrimony, clearing the land of its people and leasing it to sheep farmers.

Sheep were thought to be much more profitable than people and 1792 became notorious as *Am Blaidhna nam Caorach Mora*, 'The Year of the Big Sheep'. Imported from the south (and often accompanied by experienced Border shepherds), large flocks of Cheviot sheep were hefted to Highland pasture. Tough and much bigger than the native breeds, they thrived, especially in the cleared crofting townships of Sutherland and Ross. The people, however, did not fare so well.

Between 1811 and 1820, a brutal aristocracy did not hesitate and evictions were forced at an astonishing speed. More than 500 families were turned out of their houses in one day. It was pitiless, callous. In a letter to friends in England, Elizabeth Leveson-Gower, the Duchess of Sutherland, wrote about the starving crofters on her estate: 'Scotch people are of happier constitution and do not fatten like the larger breed of animals.'

One of the lean – but eloquent – crofters, Donald McLeod, wrote an account of what he himself saw. By 'terrific' he meant terrifying.

The consternation and confusion were extreme. Little or no time was given for the removal of persons or property; the people striving to remove the sick and the helpless before the fire should reach them; next struggling to save the most valuable of their effects. The cries of the women and children, the roaring of the affrighted cattle, hunted at the same time by the yelling dogs of the shepherds amid the smoke and fire, altogether presented a scene that completely baffles description – it required to be seen to be believed.

A dense cloud of smoke enveloped the whole country by day,

and even extended far out to sea. At night an awfully grand but terrific scene presented itself – all the houses in an extensive district in flames at once. I myself ascended a height at about eleven o'clock in the evening, and counted two hundred and fifty blazing houses, many of the owners of which I personally knew, but whose present condition – whether in or out of the flames – I could not tell. The conflagration lasted six days, till the whole of the dwellings were reduced to ashes or smoking ruins. During one of those days a boat actually lost her way in the dense smoke as she approached the shore, but at night was enabled to reach a landing-place by the lurid light of the flames.

The ships that found their way to shore and the emigrants waiting at the landing places took them away from all that shock and devastation and scattered them to the corners of the earth. In the USA, Canada, Australia and New Zealand, there are many more people of Highland ancestry than now live in the Highlands.

Even after the War of Independence of 1775–83, most emigrants sailed to the United States – perhaps precisely because it was independent of Great Britain. Between 1815 and 1914, more than 13 million Scots arrived in the USA (four million went to Canada and a million and a half to Australia) and these settlers were very influential. The census of 1790 showed that 12 per cent of the new nation was of Scots or 'Scotch-Irish' descent. The latter were also known as Ulster Scots and were the descendants of communities planted in Northern Ireland, mostly from the Scottish Borders and the Lowlands. The fifth American President, James Monroe, was the direct descendant of a minor clan chief and the seventh, Andrew Jackson, hailed from hardy Scotch-Irish stock. In all, a staggering 23 of the USA's presidents, more than half, have had Scots or Scotch-Irish lineages in their family tree.

This statistic is all the more remarkable when thought of in terms of the background of mass emigration. As a result of the arrival of many more ethnic groups and the effects of slavery, the proportion of people claiming Scots descent in the USA has declined to 1.7 per cent. The very definition of a melting pot, the dynamism of American demographics inevitably meant a dilution and mixture of the DNA of early incoming peoples. For example, 30 per cent of the Y chromosomes of African-Americans are European, reflecting the legacy of the slave generations. A considerable proportion of these are likely to have been Scots in origin. Amongst the Cree people of Canada, there are Orcadian surnames such as Linklater, Flett and Foubister. Many men from Orkney worked for the Hudson's Bay Company. Despite the complex genetic mixture in all of these former British colonies, there can be no doubt that Scottish lineages which are now extinct at home carry on in the USA, Canada, Australia and New Zealand. Surname evidence alone shows this to be the case. Whatever may happen in Scotland, Scottish DNA is still on its journey to the future.

In Scotland, it is changing – our collective DNA has been enriched and somewhat altered in the recent past by a series of immigrations of significant scale and character. After centuries of population loss, the trend has reversed, and since the late 1980s there has been a net migration gain. The Scottish population rose from an estimated 5,083,000 in 1991 to an estimated 5,194,000 in 2009. The census asks respondents to define their own ethnicity and the overwhelming mass reckon themselves White Europeans at 98.19 per cent and, of these, 88.09 per cent see themselves as Scots. Other White British, mostly English immigrants, account for 7.38 per cent, and Other Whites and Irish people make up the balance at 2.71 per cent. But inside these blunt numbers are some fascinating stories.

HOW MANY AND WHO

Population of Scotland estimated in 2009: 5,194,000
Ethnic groups

White	4,960,334 (98.19%)
Scottish	4,459,071 (88.09%)
Other White British	373,685 (7.38%)
Any other White background	87,650 (1.73%)
White Irish	49,248 (0.98%)
Mixed	12,764 (0.25%)

South Asian	55,007 (1.09%)
Pakistani	31,793 (0.63%)
Indian	15,037 (0.30%)
Bangladeshi	1,981 (0.04%)
Other South Asian	6,196 (0.12%)
Black	8,025 (0.16%)
African	5,118 (0.10%)
Caribbean	1,778 (0.04%)
Other Black	1,129 (0.20%)
Chinese	16,310 (0.32%)

Emigration from Ireland to Scotland did not cease with the arrival of the descendants of Niall Noigiallach and their compatriots. Sea travel across the North Channel was frequent and easy. A deck passage for the short voyage between Ireland and Greenock cost only sixpence and substantial seasonal waves of immigration took place every year. Within living memory, squads of Irish 'tattie howkers' came to work in Scotland at harvest time, earning cash wages and sleeping in very basic bothy accommodation or even in hay barns. This pattern was well established as early as the 1820s when 6,000 to 8,000 Irish labourers came across for the harvest season, beginning with soft fruit and cereals in late August and ending with potatoes in October. Numbers increased steadily and, by 1841, almost 58,000 temporary workers from Ireland were living in Britain. The economy was booming and railway construction drew many gangs of Irish navvies (a shortened version of 'navigator', it was a term coined for the men who dug Britain's canal network in the late eighteenth century) to dig the cuttings and tunnels and pile up the embankments needed to carry the trackbed. In an age before earth-moving equipment, a navvy's shovel and barrow shifted millions of tons of soil and stone.

These men lived in notoriously squalid temporary accommodation. Some who worked on the railways of southern Scotland were in the habit of digging shelters into embankments or taking refuge from the elements in shored-up tunnel entrances. It was a very hard life and many blew their cash on weekend drinking bouts that sometimes ended in violence. In areas where railways were being laid, local newspapers often carried advertisements warning people of the date navvies were due to be paid. Nevertheless the press refused to stir up resentment. Here is the measured and liberal reaction of the leader writer in *The Glasgow Courier* in 1830:

> In our opinion, the Irish have as much right to come to this country to better their lives as the Scots and English have to go to Ireland or any other parts of Britain . . . Let us hear no more complaints about the influx of Irish . . .

Two years earlier the sensational trial of two former Irish navvies, William Burke and William Hare, failed to persuade the volatile Edinburgh mob to turn out and give vent to anti-Irish sentiment. Despite their high visibility and mobility, the immigrant workers produced the opposite of animosity from the other end of Edinburgh's social scale. Here is an entry from the diary of Henry Cockburn, the judge and littérateur:

> The whole country was overrun by Irish labourers, so that the Presbyterian population learned experimentally that a man might be a Catholic without having the passions or the visible horns of the Devil. New chapels have arisen peaceably everywhere; and except their stronger taste for a fight now and then, the Irish have in many places behaved fully as well as our own people. The recent extinction of civil disability on account of the religion removed the legal encouragement of intolerance, and left common-sense some chance; and the mere habit of hating, and of thinking it a duty to act on this feeling, being superseded, Catholics and rational Protestants are more friendly than the different sects of Protestants are.

The 1841 census counted 126,321 Irish-born people in Scotland, around 5 per cent of the population. Almost a third settled in Glasgow and found work in the growing heavy industries of Clydeside. Many were dockers or labourers in the huge foundries at Parkhead Forge or Beardmore's. With the devastating potato famine of 1845–51, when more than a million Irish people died, emigration accelerated. Of the three million who left, 75,000 came to Scotland. At its peak in 1848, the average number of weekly immigrants disembarking at the quays in Glasgow was estimated at more than a thousand but, between January and April of that year, 42,860 came from Ireland. Even in a city as large as Glasgow and one used to spurts of spectacular growth, this influx must have made an immediate impact. It was probably the most intense episode of immigration into Scotland for a thousand years or more and the visibility of the incoming Irish settlers was greatly magnified by their concentration in Glasgow and North Lanarkshire. Coal mining and the heavy industries clustering around the coalfields were growing fast, and in the factories and forges of Airdrie, Coatbridge and Motherwell, many immigrants quickly found work. It was hard, menial and often frustrating. Many Irish Catholics found it difficult to rise into the ranks of skilled workers because of religious and racial prejudice but the steel makers and manufacturers paid wages and men could put bread on the table and a roof over the heads of their families. After the horrors of the Great Famine, arrival in industrial Scotland must have seemed to many like a deliverance.

Not all immigrants disappeared into the smoke and soot of North Lanarkshire. Poverty persuaded many to take the shortest passage from Ireland to Scotland. By 1841, Wigtownshire and Kirkcudbrightshire had large Irish populations and almost all were farm workers. These jobs needed the kinds of skills the immigrants already had and, as settlers, they moved from being seasonal harvesters to more permanent jobs, often with accommodation attached. To feed the growing towns and cities to the north, farming in Galloway was also expanding, trying new and better methods to drive up yields. By 1851, Wigtownshire had a large Irish-born population at 16.5 per cent of all the people living in the county.

The overwhelming majority of the immigrants came from Ulster and their DNA echoed that of the Dalriadans 15 centuries before. While M222 and other old markers will have paid the sixpenny passage to Scotland, their addition does not confuse the overall historical picture. The great immigrations of the nineteenth century lie within family memory and most of the descendants of those who left Ulster will know something of their origins.

Around 75–80 per cent of the new arrivals were Catholic. The rest were Protestants who had also been driven by poverty and a lack of opportunity to seek new lives. Old hatreds formed part of the baggage of the immigrants as the orange and green divide took shape, especially in North Lanarkshire, Ayrshire and Glasgow. Many Orange Lodges were founded and annual parades began to be seen on 12 July, the commemoration of the Battle of the Boyne.

The battle lines of sectarianism mustered in Scotland. Protestant Ulstermen tended to monopolise the better-paid and more highly skilled jobs while Catholics set about creating their own institutions. Chapels and Catholic schools were funded and built and the community turned in on itself somewhat. In 1851 in Greenock, records show that 80.6 per cent of Irish-born men and women found marriage partners amongst the existing Irish community.

The surge in immigration after the Great Famine not only drove up the percentage of Irish-born to 7.2, it also fuelled resentment and religious bigotry. Fearing the growing power of the Catholic Church, the Church of Scotland attacked immigration but the Irish communities grew strong, well organised and settled. Dundee was exceptional in welcoming Irish women to work in the expanding jute industry. Hibernian Football Club came into being in Edinburgh in 1875 and 13 years later Celtic Football Club began playing matches in Glasgow.

Bigotry had deep and durable roots and as recently as 1923 the Church of Scotland published a pamphlet in which it described its anxiety about 'the menace of the Irish race to our Scottish nationality'. When, in the late 1920s, the Great Depression began to bite hard, causing widespread unemployment, two anti-Irish organisations sprang up – the Scottish Protestant League in Glasgow and Protestant Action in Edinburgh. Under the leadership of John Cormack, the latter incited a mob of more than 10,000 to attack participants in the Catholic Eucharistic Conference which was being held in the city in 1935. Buses carrying children were stoned and Catholics organised all-night vigils to protect their churches from vandalism. The Second World War put an abrupt end to these ugly incidents, and since 1945 the most visible spur to sectarian behaviour has been the football teams supported by opposing factions.

When immigrants arrive in groups, their instinct is for cohesion, a cooperative effort to make sense of and survive in a new environment. Everything is foreign to them and they are foreign to the natives. The first Jewish congregation gathered for worship in Edinburgh in 1816 and a second synagogue was founded in Glasgow in 1823. During the nineteenth century, more Jews arrived, many fleeing from persecution in the Russian Empire.

More than five million Jews lived in a large area known as the Pale. It comprised much of Lithuania, Poland, Belarus, Moldova, Ukraine and western Russia. Established in 1791 by Catherine the Great, it was intended as a clearly demarcated region where Jews were permitted to live and work – and be taxed. Exceptionally, some were given dispensations to reside elsewhere but the Russian Empress wished to contain all Jews in one sector of her vast dominions. Part of her motivation was to allow the rise of a native Russian middle class, people who did not fit into the ancient and rigid hierarchies of aristocrats, serfs and clerics. Catherine wished to encourage a new group whose dynamism might modernise and create wealth.

As the greatest concentration of Jews anywhere, the Pale of Settlement was also a target. Anti-semitic violence broke out regularly as mobs rampaged through the Jewish towns known as *shtetls*. But, until the late nineteenth century, there were also periods of relative peace. For religious and cultural reasons, as well as the instinct for mutual support and protection, *shtetls* were mostly inward-looking and conservative. The Barbra Streisand film, *Yentl*, in which she played a young tomboy who wanted to enter religious training, something usually reserved for men, was seen in its context as very shocking – although it did give a strong sense of what life was like in the *shtetls*.

Bouts of anti-semitic rioting and violence known as pogroms began in earnest in 1881 and for two years all across the Pale, Jewish houses and businesses were burned. Many were killed. The pogroms sparked immediate emigration and many Jews fled the Pale forever. Because their nearest points of embarkation were the Baltic ports, families and individuals found themselves landing at Scottish and eastern English quaysides. Most sought sanctuary in *die goldene medine,* 'the Golden Land' of the United States, and Scotland was only a staging post. But some got no further and a sizeable Jewish community grew up, especially in Glasgow.

Unlike the much earlier immigrant conquerors who crossed the North Sea and the North Channel many centuries before, the Jews (and the recently arrived Irish) found themselves at the very bottom of the social scale. Many Jews settled in the Gorbals, a district of high population density on the banks of the Clyde, where tenements were packed with families living in slum conditions. But no one persecuted the new arrivals and, while there was some prejudice, blood-thirsty mobs were not about to attack and burn the Gorbals to the ground.

In a pleasing irony, the only political institution committed to a policy of anti-semitism, the British Union of Fascists, were not welcomed in Scotland – but for all the wrong reasons. When the BUF leader, Oswald Mosley, swaggered into Edinburgh in 1934, he and his followers were attacked on Princes Street by the Protestant Action group. John Cormack believed that the fascists were Italians and therefore dangerous Roman Catholics.

From the *shtetls,* Jews had brought skills and some quickly found niches in the local economy. Cigarette and cigar making was one specialism, while other families began to create retail businesses that would grow and flourish into the twentieth century. Some historians are of the view that Jewish self-betterment was not hindered in Scotland because bigotry was obsessed with the imagined threats of Irish Catholicism.

In 1905, an unprecedented 8,000 Jews arrived in Scotland from Eastern Europe. They were seeking sanctuary from a new outbreak of pogroms and government-sponsored oppression. The plot of the famous musical, *Fiddler on the Roof,* turns on the Tsarist decree evicting people from their *shtetl.* The Jewish economist and writer Ralph Glasser was born in the Gorbals and he remembered the aftermath of the refugees coming to Scotland from the *shtetls*:

The new arrival was quickly spotted. A man with a week's growth of beard, eyes bleary from wakefulness in his long journey, would shuffle wearily through Gorbals Street with his 'peckel', his belongings, strapped in a misshapen suitcase, listening for the familiar tones of this lingua-Judaica from the East European Marches, and approach such a group with the sureness of a questing bloodhound. He would fumble in the pocket of a shapeless coat and show them a much-thumbed envelope.

'Lansmann! Sogmer, wo traffic dos?' ('Fellow countryman. Tell me, where can I find this address/person?').

This example shows step-migration in action, with those who arrive first in a new place bringing others in their wake. As the cancer of anti-semitism grew in the first half of the twentieth century, many more Jews came from Eastern Europe. By 1950, the Jewish community in Scotland had grown to 80,000. The writer and academic David Daiches believed Scotland to be a haven when he noted that, of all the countries of Western Europe, it had no history of institutional anti-semitism.

Now the Jewish community is much reduced. Glasgow was its centre for many years, but according to the 2001 census only about 5,000 live in and around the city. Most of the remainder of a tiny remnant of 6,400 live in Edinburgh and Dundee. The reasons for decline are straightforward. When individuals marry outside their faith (and Jewish custom is clear about this), they usually cease to be formally Jewish. After integration, emigration has caused the Scottish community to shrink. Many have gone south to Manchester or London, some to settle in Israel.

Genetically the numbers are too transient and small to be significant but the cultural contribution of Jews to Scotland was and continues to be immense and disproportionate.

An attractive footnote to Scots-Jewish history is the story of the sons of Aaron. Some men carrying what is known as the Cohen Modal Haplotype will be found in Scotland. This is a Y chromosome in the P58 group which is highly enriched amongst the Cohanim, the Jewish High Priests, and which confirms that a very large proportion descend from one man who lived about 2,000 years ago. According to oral traditions, this may have been Aaron, Moses's brother.

Some recent Eastern European immigrants saw their identity

submerge very quickly. Between the 1860s and 1914, 650,000 Lithuanians left their Baltic homeland, a quarter of the entire population. An aggressive programme of Russification threatened many of their most distinctive characteristics. Lithuanian Catholicism was to be replaced by Russian Orthodox rites, young men were to be conscripted into the Tsarist army and their agrarian economy was seriously disrupted. When the Tsar emancipated Russia's serfs, this seemingly enlightened measure had the effect of driving up the rents and taxes to be paid by Lithuanian farmers.

Scottish coal-owners and steel makers such as Dixons and Bairds had, in any case, been recruiting Lithuanians and when persecution drove many to board ships sailing to the west and relative freedom, employers quickly found jobs for the new immigrants. Settlement concentrated in North Lanarkshire, especially Bellshill and Coatbridge, and in Ayrshire, Fife and West Lothian. At first, trade unionists were antagonistic to men brought in to undercut their members' wages and even to break strikes but the incomers soon joined unions and subscribed to their social and political aims.

The early settlers enthusiastically maintained their Lithuanian culture. Wedding celebrations took a week, christenings three days, plum brandy was drunk, Eastern European foods like sausage and rye bread were made and the devoutly Catholic community celebrated all the major festivals. But over the span of only two generations, assimilation began. In the jobless decade of the Great Depression, many Lithuanians changed their names, sometimes arbitrarily to Smith or Black, because they reckoned it might help in the search for employment. Many did not wish to employ people with alien names or Catholics. In contrast with the Irish and Jewish communities, there were no Lithuanian schools and English quickly replaced the language of their homeland. At only 8,000 in 1914, the Lithuanians in Scotland were a small group and they could be absorbed easily and quietly in the large towns and cities. Few traces of this colourful and initially vibrant community now remain.

It was agricultural poverty rather than persecution that drove Italians to settle in Scotland. In the 1890s, families left the south of Italy (some of them walked all the way) in particular to escape famine and drought. In a classic process of chain migration, Italians who became established often welcomed more incomers, usually family or people from the same community. Two major sources for immigrants were the towns of Barga in northern Tuscany, near Lucca, and Frosinone in Lazio, south of Rome. Chain migration was much encouraged by the *padrone* system whereby Italians established in Scotland paid for the passage of young men from Italy and gave them a job. Usually the *padroni* were businessmen anxious to expand and to recruit Italians, often from the same locality as themselves. They hoped that they would work cheaply and work hard, just as they themselves had done.

By 1914, there were 4,500 Italian-born immigrants in Scotland, many of them working in the food industry. Ice cream and fish and chips were the staples created and marketed by enterprising families. At first ice cream was sold in the streets from barrows or carts. Much cheaper than renting or buying a shop, it was an occupation that gave many a start in business. Italians invented the 'pokey hat' or ice-cream cornet and it made the whole retail operation simple in that everything could be eaten, including the receptacle for the ice cream. Fish and chips was the original fast food for ordinary people and it used two ingredients very readily available in Scotland. It appears to have been pioneered in East London by Jewish families but it was certainly made popular and widely available in Scotland by Italians. It too was street food in that it was originally sold to be eaten in the open air and visits to the chippy after the pubs closed at least soaked up some of the beer.

Italians also began to establish cafes and, since these stayed open into the evenings, as they did in Italy, and much longer than other similar establishments, they quickly became busy. The more straight-laced disapproved of young people of both sexes spending time together unsupervised in the evening but the temperance societies approved – Italian cafes sold coffee, tea and soft drinks only.

Italian involvement in the food industry meant that the community had to disperse. Unlike other groups of immigrants who tended to concentrate in particular places, the Italians were forced to settle all over Scotland to found their businesses and avoid competing with each other. Soon almost every small town and city district had its ice-cream parlour, cafe and chip shop, often all run from the same premises. This meant a widespread familiarity with Italian-Scots and they were popular – at least at first.

Mussolini's government made a point of reaching out to the Italian diaspora. The first fascist club in Scotland was founded in Glasgow in 1922 and several others quickly followed in Scotland's other cities. Many Italian-Scots, perhaps 50 per cent, were members of the Fascist Party although their affinity appears to have been more patriotic than political. When Mussolini declared war on Britain in 1940, there was an immediate backlash. Rioting crowds attacked and damaged Italian shops and businesses and all Italian men between the ages of 17 and 60 were arrested and interned. Many were transported to Canada, Australia and elsewhere. In 1940, the *Arandora Star* was carrying internees to Canada when she was sunk by a German U-boat. The loss of life was severe and 450 Italian internees drowned. For some time after the war, there was bitterness on both sides.

Particularly in small towns where Italian families were part of the social fabric, many Scots were unhappy at their friends and neighbours being locked up. And it seems that the unhappy experiences of the Second World War encouraged assimilation.

The cheaper air travel of recent times has meant closer contact with Italy for both immigrants and their descendants – and for native Scots. One of the effects of better cultural understanding has been to drive Italian-Scottish businesses up market. The wonderful Edinburgh grocery and wine merchant, Valvona & Crolla, began life catering for purely Italian tastes but its customer base is now international.

Recent research reckoned the Scots-Italians at between 70,000 and 100,000. Since DNA around the shores of the Mediterranean is more diverse than in north-western Europe, a group of that size is likely to show up in any collective genetic map of Scotland. Glasgow hosts the third largest Italian community in Britain and their presence is enhanced by a continuing involvement in the food industry with chip shops, ice-cream parlours, delicatessens and restaurants on many high streets.

A remarkable case study showed how rich the Italian contribution could be in Scotland, and the links it could make. The distinguished actor Tom Conti took an ancestral DNA test. He turned out to be in the Y chromosome haplogroup M34, subtitled Saracen. It may have come to the Tyrrhenian coast of Italy with the Saracen pirates who terrorised communities from the eighth century onwards, and Tom inherited it from his father, Alfonso, who came to settle in Scotland in the 1930s – from a town on the Tyrrhenian coast.

At around the same time genetic research into the ancestry of a very famous Frenchman was going on in Paris. Some of Napoleon Bonaparte's beard hairs had been preserved in a reliquary and DNA was extracted from their roots. The genetic marker from it proved to be identical to the one found in Prince Charles Napoleon Bonaparte, the French politician who is the great-great-grandnephew of Napoleon, a descendant in the direct male line. All have an M34 marker, including Tom Conti. He is clearly a distant relative of Napoleon since they share a recent common ancestor. Conti was delighted.

Poland Street in the Soho district of London was home to an early concentration of exiles. When the November Uprising of 1831 against the Russian occupation failed, many Poles came to Britain to seek political asylum. Immigration to Scotland in large numbers only began after World War Two. After the Nazi and Soviet occupation of Poland in 1939, many political émigrés fled west and they were accompanied by more than 20,000 soldiers and airmen when they arrived in Britain. By 1945, there were 228,000 serving in the Polish Armed Forces in the West, the fourth-largest armed force fighting alongside the British, the Americans and the Russians. They made an immense contribution, most notably in the Battle of Britain and in cracking an early version of the Enigma codes.

The Polish soldiers who joined the Allied armies had suffered terrible privations on their long journey to the Normandy beaches and the ultimate liberation of Western Europe. Many came from the Kresy region of eastern Poland, in particular the cities of Lwow and Wilno, and when the Nazis and the Russians occupied their country in 1939, they were captured and sent to the gulags. When Stalin was forced to change sides in 1941 and join the Allies in the war against Hitler, the imprisoned Kresy Poles were released and, under the command of General Anders, they marched across Siberia to Persia to link with the Allied Armies in the Middle East. After playing a central role in the defeat of the Germans in North Africa and Italy, where their courage under fire became legendary, the men of the Anders army hoped eventually to return home. But when the Russians were allowed to retain those parts of Poland they occupied in 1939, many tens of thousands of men were devastated, knowing that if they went back to Kresy they would be persecuted and that if they did not their families might suffer.

When the dismal details of the political deal made by Churchill, Roosevelt and Stalin became clear, the despair of some Polish soldiers became unbearable. Having fought their way through some of the bloodiest battles of the war, 30 officers and men of II Corps committed suicide. As inadequate recompense for those who remained, as another blood price, the British government offered citizenship and the 1947 Polish Resettlement Act recognised a community of 162,000.

When communism at last fell and a freely elected President of Poland took office in 1991, the exiled government in London was dissolved. For old men with long memories, the march through the wastes of Siberia had finally ended.

In 2004 the European Union expanded and the United Kingdom granted free movement to workers from the new member states, including Poland. Possibly in part stimulated by a preexisting series of personal and cultural links, many came to work in Britain and the size of the Polish community has risen steeply. The Office for National Statistics estimated it at 520,000 in 2009 and some believe it is much higher, perhaps a million strong.

In Scotland the old post-war Polish communities had largely integrated by the close of the twentieth century, their social clubs and shops less frequented by second and third generations, but the new influx has reinvigorated a Polish presence. Now there are thought to be between 40,000 and 50,000 people living in Scotland who were born in Poland and a bilingual Polish–English newspaper is widely read.

DNA links the Poles closely to the final, visible new group in Scotland, the Pakistanis. Surprisingly both Poles and Pakistanis share a very high frequency of the M17 marker with Norwegians of Viking descent. These links were made in the deep past but they are, nevertheless, there, producing one more quirky connection. A Viking descendant in Orkney, Shetland, Caithness or the Western Isles is likely to be more closely related to a Pakistani or a Pole than he is to other Scots in the male line. In the small South Asian community of 55,000, Pakistanis dominate with 31,793, more than double the number of Indians at 15,037. In all, South Asians make up 1 per cent of the Scottish population but their visibility on the high streets and in the catering industry belies small numbers.

The largest immigrant group in Scotland in modern times is also the hardest to detect. The number of English-born people living in Scotland has risen markedly since 1841 when it stood at 1.5 per cent. According to the estimates of the General Register Office of Scotland, in 2006 there were 373,685 English men and women resident, 7.38 per cent of the population.

In terms of their DNA, the English and the Scots share a great deal. But attitudes differ. Many Scots welcome people who actively choose to live in Scotland, who made a conscious decision to come north. Others resent those known as incomers, white settlers, interlopers and a row of other denigrating terms. In his novel *Trainspotting*, Irvine Welsh put these paradoxical thoughts in the mind of his hero, Renton:

> It's nae good blaming it oan the English for colonising us. Ah don't hate the English. They're just wankers. We are colonised by wankers. We can't even pick a decent, vibrant, healthy culture to be colonised by. No. We're ruled by effete arseholes. What does that make us? . . . The most wretched, servile, miserable, pathetic trash that was ever shat intae creation.

Or not. While that jumble of competing emotions may be very readily recognisable in our dealings with England and the English, there are more measured views on record. In his fascinating study *Being English in Scotland*, Murray Watson interviewed many for their views and attempted to fill what he saw as a historiographical void

– the lack of a decent, extended treatment of English immigration to Scotland. Here is his conclusion:

> Generally, throughout the period under review [broadly, the modern period], the media painted a picture of a climate of anti-English feeling. This was not the general experience of the contributors, nor was it evident from other sources. Studies from a number of social scientists, albeit they were mostly restricted to peripheral areas, essentially corroborated the findings of this study. That was not to say that tensions did not exist. There were low levels of anti-English feeling and exceptional extremist activity, but the latter was largely directed against England, the state, and not English people. Compared with prejudicial reactions to other migrant communities, the English were largely welcomed into Scottish society, and this is certainly borne out by the constant growth of English migrants settling in Scotland.

> This expansion in numbers came at a time of decline and stagnation in Scotland's population. At the outset of the twenty-first century civil servants and politicians have been calling for immigration policies to arrest the downturn. And as the nationalist MSP, English-born Mike Russell, put it, 'Scotland is not full up.' Throughout the second half of the twentieth century and in the early twenty-first century the English, an invisible diaspora, have played a significant role in Scotland's demography, society, culture, economy and politics.

The national experience of immigration and emigration in the eighteenth, nineteenth and twentieth centuries has shown something simple – and cheering. DNA is dynamic, always changing, mutating, often producing the unexpected, making a nonsense of stereotypes. The genetic journey of the Scots is not over but is still in train. Over time the pace slackens, accelerates, appears to stall then is jolted unexpectedly into motion. It never stops. The answer to the child's question about where they come from turns out to have many colourful, nuanced and provisional responses. Scottishness has no set of standard elements; our genetic journeys show that it glints with many facets. We are all immigrants and we are all Scots.

Bibliography

This is a general reading list for those who may wish to pursue their own course of study. It omits the many learned academic articles available to more specialised interests and also the vast resource that is the Internet.

Bahn, Paul G., *Cave Art*, Frances Lincoln, 2007
Barton, Nick, *Ice Age Britain*, Batsford, 1997
Burley, Robbins, *The Talking Ape*, OUP, 2005
Clarkson, Tim, *The Men of the North*, John Donald, 2010
Cunliffe, Barry, *The Ancient Celts*, Penguin, 1997
Cunliffe, Barry, *Europe Between the Oceans*, Yale UP, 2008
Cunliffe, Barry (ed.), *The Penguin Atlas of British and Irish History*, Penguin, 2001
Cunliffe, Barry (ed.), *The Oxford Illustrated History of Prehistoric Europe*, OUP, 1994
Curtis, Gregory, *The Cave Painters*, Anchor Books, 2006
Dawkins, Richard, *The Ancestor's Tale*, Orion, 2004
Deutscher, Guy, *The Unfolding of Language*, Arrow Books, 2006
Devine, Tom, *The Scottish Nation*, Penguin, 1999
Diamond, Jared, *Guns, Germs and Steel*, Norton, 1997
Fagan, Brian, *The Long Summer*, Granta, 2004
Gaffney, Vince, *Europe's Lost World*, CBA, 2009
Herodotus, *The Histories*, Penguin Classics, 1994

Bibliography

Kennedy, Maeve, *Archaeology*, Hamlyn, 1998

Lewis-Williams, David and Pearce, David, *Inside the Neolithic Mind*, Thames & Hudson, 2008

Lynch, Michael, *Scotland: A New History*, Pimlico, 1991

MacLeod, Mona, *Leaving Scotland*, National Museum of Scotland, 1996

McKie, Robin, *The Face of Britain*, Simon & Schuster, 2006

Miles, David, *The Tribes of Britain*, Weidenfeld & Nicolson, 2005

Mithen, Steven, *After the Ice*, Weidenfeld & Nicolson, 2003

Moffat, Alistair, *The Borders: A History From Earliest Times*, Birlinn, 2002

Morris, Desmond, *Horsewatching*, Ebury Press, 2000

Nicolaisen, W.F.H., *Scottish Placenames*, Batsford, 1976

Oppenheimer, Stephen, *The Origins of the British*, Robinson, 2006

Ostler, N., *Empires of the Word*, HarperCollins, 2005

The Oxford Companion to the Earth, OUP, 2000

Pitts, Mike, *Hengeworld*, Arrow Books, 2001

Ralston, Ian (ed.), *The Archaeology of Britain*, Routledge, 1999

Renfrew, Colin, *Prehistory*, Weidenfeld & Nicolson, 2007

Smyth, A.P., *Warlords and Holy Men*, EUP, 1984

Stringer, Chris, *Homo Britannicus*, Penguin, 2006

Sykes, Bryan, *The Blood of the Isles*, Bantam Press, 2006

Sykes, Bryan, *The Seven Daughters of Eve*, Bantam Press, 2001

Index

✶

Index

Index

Index